T0310594

LOGIC, PROBABILITY AND SCIENCE

POZNAŃ STUDIES
IN THE PHILOSOPHY OF THE SCIENCES AND THE HUMANITIES

VOLUME 71

EDITORS

Krzysztof Brzechczyn (assistant editor)
Jerzy Brzeziński
Michał Cichocki (assistant editor)
Andrzej Klawiter
Piotr Kwieciński (assistant editor)

Krzysztof Łastowski
Leszek Nowak (editor-in-chief)
Izabella Nowakowa
Katarzyna Paprzycka (Pittsburgh)
Marcin Paprzycki (Hattiesburg)
Piotr Przybysz (assistant editor)

ADVISORY COMMITTEE

Joseph Agassi (Tel-Aviv)
Étienne Balibar (Paris)
Wolfgang Balzer (München)
Mario Bunge (Montreal)
Nancy Cartwright (London)
Robert S. Cohen (Boston)
Francesco Coniglione (Catania)
Andrzej Falkiewicz (Wrocław)
Dagfinn Føllesdal (Oslo)
Bert Hamminga (Tilburg)
Jaakko Hintikka (Boston)
Jacek J. Jadacki (Warszawa)
Jerzy Kmita (Poznań)

Leon Koj (Lublin)
Władysław Krajewski (Warszawa)
Theo A.F. Kuipers (Groningen)
Witold Marciszewski (Warszawa)
Ilkka Niiniluoto (Helsinki)
Günter Patzig (Göttingen)
Jerzy Perzanowski (Toruń)
Marian Przełęcki (Warszawa)
Jan Such (Poznań)
Max Urchs (Konstanz)
Jan Woleński (Kraków)
Ryszard Wójcicki (Warszawa)
Georg H. von Wright (Helsinki)

This book series is partly sponsored by Adam Mickiewicz University, Poznań

The address: prof. L. Nowak, Cybulskiego 13, 60-247 Poznań, Poland.
Fax: (061) 8477-079 or (061) 8471-555
E-mail: epistemo@main.amu.edu.pl

LOGIC,
PROBABILITY
AND SCIENCE

Edited by

Niall Shanks
Robert B. Gardner

Amsterdam – Atlanta, GA 2000

The paper on which this book is printed meets the requirements of "ISO 9706:1994, Information and documentation - Paper for documents - Requirements for permanence".

ISSN 0303-8157
ISBN: 90-420-1253-6 (bound)
©Editions Rodopi B.V., Amsterdam - Atlanta, GA 2000
Printed in The Netherlands

In memory of
Amy Weems

CONTENTS

Introduction

The essays gathered in this volume explore philosophical issues in logic, probability theory and the natural sciences. Though the volume explores diverse problems in several different fields of inquiry, a common thread running through the essays is that philosophical insight is brought about through the application of rigorous analytical methods to problems of interest. The essays printed here belong firmly in the broad tradition of analytic philosophy. The conceptual virtues they celebrate are exactitude and clarity.

The volume grew out of papers that were presented at the annual meeting of the Society for Exact Philosophy, held at East Tennessee State University in October 1996. The editors have also included three essays that were not presented at the conference, but which are embedded firmly in the tradition of exact philosophy. Our volume should not be viewed merely as the proceedings of a conference. The authors of the essays were given time to revise their papers in the light of discussions and analyzes of their work at the conference. These revised essays were then sent to commentators, whose comments are also published here, along with replies and reactions from authors of the essays.

The first two essays concern probabilistic logic and semantics. In *Canonical Models and Probabilistic Semantics*, Charles Morgan develops the concept of canonical probability distributions in probabilistic semantics by analogy with the concept of canonical models in modal logics. Morgan aims to prove that for (almost) every extension of classical sentence logic there is a characteristic probabilistic semantics with canonical probability distributions. In the course of his proof, Morgan challenges the usual assumption that probability functions are *a priori*, one-place functions. It is Morgan's contention that even in simple probabilistic contexts, background assumptions (concerning, for example, shapes of dice, colors of balls in an urn, and so on) play a fundamental role. To accommodate this feature of these examples, Morgan treats probability functions as two-place functions. As Morgan notes in his conclusion, his approach applies to a very general class of logics and he speculates that his characterization of default maximally consistent extensions may prove of value in the treatment of counterfactuals and general default logics.

In *A Many-Valued Probabilistic Conditional Logic*, François Lepage discusses some issues arising out of David Lewis' famous paper, "Probability of Conditionals and Conditional Probabilities." To evade some unintuitive consequences of Stalnaker's treatment of counterfactual conditionals, Lewis offered his System of Spheres Semantics (*SOS*). Here, possible worlds are not linearly ordered, but weakly ordered, so many possible worlds may be the same distance from a given world. As Lepage puts it, "The best image is of embedded spheres of possible worlds centered on the world of evaluation. All worlds of a given layer are equidistant from the world of evaluation" (p. 37). But the imaging constraint, according to which, "the probability of any proposition A is the sum of the probability of the A-worlds" (p. 37), is not compatible with Lewis's *SOS* semantics.

Lepage explores the possibility that imaging can be introduced into *SOS* by changing the requirements on the truth conditions for conditional statements. Lepage considers the possibility that under certain circumstances conditionals can take fractional truth values. In the course of his essay, he goes on to present a non-extensional logic of conditionals in which (a) all instances of tautologies are valid, and (b) if all the sub-expressions of a proposition have classical truth values, then the proposition also has a classical truth value.

The next group of three essays are broadly concerned with issues arising out of decision theory, and Bayesian approaches to epistemology. In *The Exchange Paradox, Finite Additivity, and the Principle of Dominance*, Piers Rawling analyzes the exchange paradox, otherwise known as the two envelopes problem. The problem may be stated as follows: Ten dollars have been placed in an envelope, O, and a fair coin has been tossed. If it came up heads, twenty dollars was placed in a second envelope, T. If it came up tails then five dollars was placed in T. You are given one of the envelopes, and you are given the opportunity to trade the envelope you have for the other one. The puzzle is that the expected actuarial value of T is 1.25 times that of O. But the expected actuarial value of O is also 1.25 times that of T. Trading T for O is advantageous, as is trading O for T.

Rawling reviews previous results in the literature, providing some simpler demonstrations. It is Rawling's contention that the reasoning in the traditional version of the exchange paradox is fallacious. However, he proceeds to argue that there is a variant of this problem that raises much more troubling issues. In the course of his analysis, Rawling discusses connections between the St. Petersburg paradox. He notes, "Due to their common concern with infinite expected utilities, both the St. Petersburg paradox and the two envelopes 'trading paradox' (under countable additivity) can be undercut by evading infinitude" (p. 65). Rawling considers the rejection of countable additivity — what happens if probabilities are only finitely additive? Rawling concludes that, on the basis of the two envelopes problem, "If probability is merely finitely additive, it seems we must abandon decision theoretic reasoning with respect to infinite partitions" (p. 69).

In *The Logical Status of Conditionalization and its Role in Confirmation*, Susan Vineberg considers the issue of the justification of the rule of conditionalization in Bayesian theories of confirmation and decision. The rule of conditionalization requires, "that an agent's new probability for A after learning E, and nothing more, should be equal to her old probability of A given E" (p. 77).

In the first part of her essay, Vineberg considers the role played by the Dutch strategy argument in the justification of the rule of conditionalization. A Dutch strategy is a betting strategy that secures a net loss for the bettor, and the Dutch strategy argument goes as follows: "If an agent's beliefs change, after learning E, by a rule other than conditionalization, then she is susceptible to a Dutch strategy, in that a bookie, who knows her degrees of confidence and her rule for updating, can devise a series of bets to be placed at different times, each of which would appear fair to the agent at the time offered, but which together guarantee her a net loss" (pp. 78–79).

In the light of Vineberg's analysis of the Dutch strategy argument, the rule of conditionalization emerges not as a requirement of rationality, but as a rule of permission. As Vineberg puts it, "there is no rational requirement that the hypothesis H be updated by conditionalizing on E, when E is learned, as opposed to giving up the prior conditional probability of H given E" (p. 82).

In the second part of her essay, Vineberg considers the implications of this reading of the conditionalization rule for Bayesian confirmation theory. In the course of her discussion, Vineberg examines arguments to the effect that the conditionalization rule can resolve issues raised in the realist/anti-realist debate in the philosophy of science. Vineberg contends that the conditionalization rule, viewed as a rule of permission, cannot provide a resolution of the realist/anti-realist debate.

Further issues touching upon the philosophy of science are raised by Deborah Mayo in her essay *Science, Error Statistics, and Arguing from Error*. Mayo examines two divergent views of the task of a theory of statistics: the evidential relation view (exemplified by various versions of Bayesian confirmation theory); and the error-statistical view (exemplified by the Neyman-Pearson approach to statistics). Mayo contends that while experimental investigators tend to follow some version of an error-statistical approach, many philosophers insist on a Bayesian approach to scientific inference. Mayo's central thesis, contrary to the Bayesian orthodoxy, is that an error-statistical approach can provide a good foundation for a philosophy of experimental inference.

Central to Mayo's thesis is the concept of severe test. As Mayo puts it: "Data *e* indicate the correctness of hypothesis *H*, to the extent that *H* passes a *severe test* with *e* . . . Hypothesis *H* passes a severe test with *e* if (a) *e* fits *H* and (b) the test procedure had a high probability of producing a result that accords less well with *H* than *e* does, if *H* were false or incorrect" (p. 99). In view of this, experimental inquiry is viewed as involving the construction and correction of models needed to systematically substantiate severe tests. In the end, and unlike the orthodox Bayesian approach to confirmation, the error statistical approach, "licenses claims about hypotheses that are and are not indicated by tests without assigning quantitative measures of support or probability to those hypotheses" (p. 105).

In *The Best is the Enemy of the Good*, Mark Lance offers some criticisms of the Bayesian approach to epistemology. Lance believes it is important to be able to answer questions about the rationality of belief revision, and contends that the Bayesian approach to epistemology prevents us from answering such questions. His argument centers on Bayesian idealizations of cognitive agents — the theory of rationality resulting from a consideration of such agents, "offers us . . . constraints compatible with obviously irrational attitudes" (p. 113).

In discussing the shortcomings of Bayesian epistemology, Lance examines the role played by Dutch Strategy arguments in the justification of the conditionalization rule. Lance takes issue with Vineberg's defense of Bayesianism. Recall that for Vineberg, the conditionalization rule is a rule of permission. So one can confront new evidence by conditionalizing, but one does not have to — one might revise instead one's prior probability assignments, especially since Bayesian epistemology allows crazy prior probability assignments. Lance wants to know what could motivate such a revision of prior assignments. As he remarks: "This is not to say that people shouldn't change crazy assignments of probabilities into non-crazy ones. One should, but the reason is that the assignment is crazy. The learning of new evidence does not give a Bayesian agent the reason to change" (p. 122). Drawing on his analysis of Bayesianism, Lance concludes his essay with a discussion of the role of idealization in philosophy.

The next group of three essays concern issues in, or arising out of, biological science, broadly construed. In *An Application of Bayes' Theorem to Population Genetics*, Robert Gardner and Michael Wooten attempt to elucidate practical applications of Bayes' Theorem in the context of an analysis of DNA fingerprint data. Gardner and Wooten attempt to use Bayes' Theorem to derive conditional probabilities — thereby avoiding the use of transition matrices — that can be used to resolve the problem of the determination of the degree of genetic relatedness between individuals, based on their phenotypes.

Having presented their formal derivations of conditional probabilities, Gardner and Wooten apply their methods to re-analyze a British immigration case, in which the deportation of a child was at stake, depending on the degree of genetic relatedness. A mother-son relationship was being alleged, an aunt-nephew relationship was suspected. The methods actually used in settling the case (resulting in the determination of a mother-son relationship) are restricted and not suitable for general use. Gardner and Wooten show how to apply their general Bayesian approach to resolve the issue.

In *Another Look at Group Selection*, Peter Johnson explores some foundational questions in theoretical biology. Though biological orthodoxy rejects the possibility of group selection — the idea that natural selection works at the level of groups, not individual organisms or selfish genes — Johnson argues that the question needs to be re-examined. In the course of his analysis, Johnson raises some important questions concerning the nature of biological individuals. His approach, while controversial, is refreshingly anti-reductionistic in spirit. It is Johnson's contention that selection can operate — depending on context — at various levels in the biological hierarchy of organization. Johnson does not see individual level selection and group selection as opposing forces, rather he suggests that we should think in terms of a superposition of selection at various levels. In discussing the possibility of group selection, Johnson makes useful reference to the recent work of David Wilson and Elliott Sober.

Johnson is also critical of attempts to reject group selection on the basis of Occam's razor. Complex biological systems, he contends, may require complex, messy explanations. His paper ends with a group-selectionist reappraisal of some human, sociobiological scenarios. If Johnson is right, these may be just as well explained as the result of co-adaptation within groups, as by the action of selfish genes.

In *Teleosemantics, Kripkenstein and Paradox*, Cory Juhl examines issues arising out of a biologically-inspired, naturalistic account of semantics, the teleological theory. According to the teleological theory, "just as our livers have proper functions (where these functions are determined by our evolutionary history), our brains or parts thereof have as their proper function to map onto states of affairs in particular ways" (p. 168).

Consider the word 'plus,' as in "Two plus two equals four." It is Juhl's contention that finite evolutionary histories, together with finite learning histories of individuals, is not sufficient, "to connect a given representing system with a unique infinitary object like the addition function . . ." (p. 169). The teleosemanticist is committed to the view that there are only finitely many past causal facts in the relevant history of an individual member of a given species. But then, only finitely many quantities are definable from finitely many past facts.

Juhl considers the possibility that 'plus' doesn't mean something infinitary at all. Perhaps the addition rule only governs some finite set of possible applications. But as Juhl points out, this is a view that will manifest many of the problems associated with finitism in the philosophy of mathematics. In the end Juhl is sceptical of the attempt to account of semantic norms in terms of causal theory.

The next two essays are concerned with epistemological issues. In *Constitutive and Epistemic Principles*, Daniel Bonevac notes that we organize our thought with the aid of general principles. There appear to be two distinct types of principle. Constitutive principles assert that, "satisfaction of the subject term is responsible for satisfaction of the predicate term. Something's being a contract . . . makes it obligatory to honor it" (p. 183). In epistemic principles, "satisfaction of the subject term is a good indicator of satisfaction of the predicate. Fungal respiratory infections are good indicators of underlying illness . . . but are not responsible for it" (p. 183).

Bonevac notes that these principles are non-extensional and have no representations in standard logical systems. Moreover, while such principles are evidently more than accidental generalizations, they are not universal, for they admit of exceptions. Bonevac's central task is to provide a logical theory for these principles, a theory which will offer an account of their mutual relationships.

In *Empiricism, Mathematical Truth and Mathematical Knowledge*, Otávio Bueno discusses a problem originally raised by Benacerraf concerning current interpretations of mathematical truth and mathematical knowledge. An adequate characterization of the former implies an inadequate characterization of the latter, and *vice versa*. Bueno attempts to outline an empiricist interpretation of mathematics in which mathematical truth and mathematical knowledge can be simultaneously and adequately characterized. His main strategy, building on the earlier work of da Costa and French, is to work with a weaker notion of truth than is usual — quasi-truth, and a more general concept of structure — partial structure.

Bueno hopes to construct a constructive empiricist philosophy of mathematics by analogy with van Fraassen's constructive empiricist philosophy of science. The principle benefit of an empiricist interpretation of mathematics is a view of mathematics that is not committed to an ontology of abstract objects, such as sets or functions. The result of Bueno's labors is a view according to which, "Mathematical knowledge based on quasi-true theories is the result of the construction of certain partial structures and the study of their extension to full ones" (p. 231).

The final two essays in the volume are concerned with ontological questions raised by modern physics. In *Coins and Electrons: A Unified Understanding of Probabilistic Objects*, Chuang Liu points out that while quantum theory is essential for an understanding of the physics of the microcosm, it provides no clear characterization of the nature of the denizens of the microcosm, for example, electrons — especially in view of the murkiness and confusion that surrounds the doctrine of wave-particle duality.

As Chuang Liu notes, electrons would not be puzzling if they could be construed as billiard balls writ small, albeit one's that are irreducibly probabilistic. "The balls in a game of bagatelle . . . are objects of this kind, and so are fair coins and dice. But the kind of probabilistic laws the electrons obey are fundamentally different from those obeyed by the classical objects . . ." (p. 244). Chuang Liu wishes to characterize this difference, but in such a way that quantum objects do not end up as being fundamentally different from classical objects — i.e., as belonging to distinct ontological categories.

Finally, Anna Maidens, in *Are Electrons Vague Objects?*, discusses the claim that vagueness resides only in language, and cannot be part of the world. The essay concerns the controversy as to whether there can be vague objects, "in the sense that identity statements involving singular terms referring to these objects might be indeterminate in truth value" (p. 261). A counterexample in the literature concerns that case of an electron captured by an atom to form an ion which then subsequently emits an electron. It has been claimed that it is ontologically indeterminate as to whether the electron captured is one and the same electron as the one emitted.

Maidens points out that we should not insist that objects are just those things for which we have definite criteria for their identity over time — as is shown by the case of the Ship of Theseus. "Similarly, because of the way electrons enter into and then emerge from entangled states there is a vagueness in the relation relating their temporal parts" (p. 272). In view of this vagueness, Maidens concludes that identity statements about these objects are similarly vague.

Niall Shanks
Dept. of Philosophy
Dept. of Biological Sciences
Dept. of Physics and Astronomy
East Tennessee State University
Johnson City, TN 36714
shanksn@etsu.edu

Robert Gardner
Dept. of Mathematics
Dept. of Physics and Astronomy
East Tennessee State University
Johnson City, TN 37614
gardnerr@etsu.edu

Acknowledgements. We would like to thank Professor Leszek Nowak, and Marcin Paprzycki and Katarzyna Paprzycka of *Poznań Studies* for their encouragement and support of this project. We would also like to thank the President and Fellows of the Institute for the Mathematical and Physical Sciences, here at ETSU, for providing facilities for manuscript preparation.

Poznań Studies in the Philosophy of the Sciences and the Humanities
2000, *vol.* 71, *pp.* 17–35

Charles Morgan

CANONICAL MODELS AND PROBABILISTIC SEMANTICS

ABSTRACT. By analogy with the notion of canonical models in modal logics, we develop the notion of canonical probability distributions in probabilistic semantics. Essentially, a canonical distribution is one in which for every non-theorem, there is some evidence on the basis of which the non-theorem is doubtful. We then prove that for (almost) every extension of classical sentence logic there is a characteristic probabilistic semantics with canonical probability distributions.

1. Motivation

For this discussion, we will consider a logic to be defined over some language $\mathcal{L}$ consisting of a denumerable set of well formed formulas (WFFs). A logic is assumed to be composed of a set of axioms and inference rules defining the proof theory, as well as a semantic component which includes a definition of the term 'model' and a mapping of WFFs into semantic values at each model. We will be more specific later in the paper. We assume familiarity with the usual formulations of classical and indexical logics.

For classical propositional logic, a model may be taken to be a set of atomic sentence letters. We may state the principle of weak completeness for classical logic as follows:

(a) Every formula which is true in all models is a theorem.

Following the Henkin style of proof, we usually proceed by considering the contrapositive:

(b) Every non-theorem is false in some model.

To prove this latter statement, we assume to be given some arbitrary non-theorem and then show that there must be some model in which the non-theorem is false. As the non-theorem is varied, we must vary the falsifying model. That is, for classical logic, there is no *single* model in which all non-theorems are false.

The situation is slightly more complicated for propositional indexical logics. For the propositional case, we may take a frame to be a non-empty set of "worlds" with an associated "accessibility relation"; a model may be taken to be a frame along with a classical model associated with each world. The principle of weak completeness may now be stated as:

(c) Every formula which is true at every world in every model is a theorem.

The Henkin style proof then leads us to consider:

(d) Every non-theorem is false at some world in some model.

As in the classical case, we may start with some arbitrary non-theorem and show in a general way that there must be a world in some model in which the non-theorem is false. This approach is nicely illustrated in Hughes and Cresswell (1968), for example.

This approach gives the impression that, as in the classical case, when we vary the non-theorem, we must also vary the model. However, this impression is mistaken for large classes of indexical logics. For certain important classes of indexical logics, it can be shown that there is one model in which all non-theorems are falsified (see Fine 1975, for details).

In the field of indexical logics, we commonly use the phrase 'canonical model' to refer to a model in which all non-theorems are falsified. The terminology seems to have originated in Segerberg (1968) and is also found in Segerberg (1971), Fine (1975) and a host of subsequent works by various authors. Although the terminology was not used, the technique of constructing a canonical model to prove completeness for indexical logics seems to have appeared first in Lemmon and Scott (1966).

The construction of a canonical model is really quite simple. For the set W of "worlds" we may use the set of all maximally consistent sets of formulas (definitions in the next section); that is, each maximally consistent set is considered to be a possible world in W. Let us use $\Box$ for the necessity operator of the object language; where Γ is any set of formulas, we define the notation '$\Box^-$' as follows:

$$\Box^-(\Gamma) = \{A : \Box A \in \Gamma\}.$$

We use this notation to define the binary accessibility relation R of the canonical model as follows:

$$\langle \Gamma, \Delta \rangle \in R \text{ iff } \Box^-(\Gamma) \subseteq \Delta.$$

Finally, to finish the canonical model construction, we associate with each world Γ the classical propositional model consisting of the set of atomic sentence letters that are members of Γ. In the usual notation using valuation functions, this assignment has the effect of defining the valuation function as follows:

$$V(A, \Gamma) = T \text{ iff } A \in \Gamma.$$

Of course the notions 'consistent' and 'maximally consistent' are relative to the particular logic under consideration. And in order to show completeness, one must prove the relevant facts about the existence of maximally consistent sets and their properties, and show that the canonical model $\langle W, R, V \rangle$ so defined satisfies the semantic restrictions appropriate for the logic.

What would correspond to the canonical model construction for probabilistic semantics? Initially we will consider this question in a very simplistic way and then move to generalize our answer.

In elementary treatments of probability, it is fashionable to treat probability functions as essentially *a priori*, i.e., as being one-place functions. We believe that such a treatment is fundamentally mistaken, as even the simplest examples imaginable require us to make background assumptions, for example assumptions about the shape of the die, the distribution of suits and denominations of the cards, the colors of the balls and cubes in the urn, and so on. So we will always treat probability functions as two-place functions. We think of the background information, or evidence, as being given by a set of sentences from the language. So a probability distribution is a mapping from $\mathcal{L} \times \mathcal{P}(\mathcal{L})$ into the closed unit interval $[1, 0]$. We read '$P(A, \Gamma) = r$' as 'the probability of A, given Γ, is r'.

Intuitively, to say that a sentence is a universal truth (is valid) should mean that it is impossible to doubt that sentence no matter what the background evidence. In other words, we say that A is probabilistically valid if and only if for every probability distribution P and every evidence set Γ, $P(A, \Gamma) = 1$. We can state the completeness theorem as follows:

(e) If for every probability distribution P and every evidence set Γ, $P(A, \Gamma) = 1$, then A is a theorem.

Again following the Henkin approach, we consider the contrapositive:

(f) If A is not a theorem, then for some probability distribution P and some evidence set Γ, $P(A, \Gamma) \neq 1$.

In modal logic, we may think of a model as being a description of one possible universe of possible worlds. In an analogous way, we may think of a binary probability distribution as a universe of (perhaps partial) possible worlds, variations in the possible worlds being picked out by changes in the evidence. So we are led to think of a canonical probability distribution P as one in which for every non-theorem A, there is some evidence Γ such that $P(A, \Gamma) \neq 1$.

So it seems that we should define a canonical probability distribution as one in which for every non-theorem, there is some reason to doubt it. We know from previous work (Morgan 1982; 1991) that for almost any extension of classical sentence logic, there is a probabilistic semantics which is characteristic for the logic. What we will show here is that for each such logic, the class of appropriate probability functions will include a canonical probability distribution, that is, a distribution in which for every non-theorem, there is some reason to doubt it.

In fact, we will be slightly more general in our treatment. Instead of just considering theorems of the logic, we could consider derivations from arbitrarily specified background assumptions. We could then define a strongly canonical probability distribution P as one such that for any WFF A and set of WFFs Γ, if A is not derivable from Γ, then there is some additional evidence Δ such that $P(A, \Gamma \cup \Delta) \neq 1$. In short, a strongly canonical probability distribution is one such that for any set of background assumptions and any non-consequence of that set, there is additional evidence that provides grounds for doubting the non-consequence. We will show that for all of the logics under consideration, the set of probability functions defined by the characteristic probabilistic semantics contains a distribution that is strongly canonical.

In most contexts, 'canonical' usually means either 'standard form' or 'law like'. The use of the phrase 'canonical model' may be appropriate in the context of modal logics in so far as there is a unique "standard form" model corresponding to each logic. However, as we shall see shortly, canonical probability distributions are not unique. That is, for a given logic, in general there will be many canonical probability distributions. We first used the term 'canonical' in this context only because of the analogy with modal logics. On the other hand, for each given logic, only the laws of that logic are immune from doubt in the canonical distributions for that logic. So while not unique, canonical distributions may be thought of as "law-like" in this sense.

2. Technical Background

Here we will lay out the specifications for the sorts of logics we wish to consider and their associated probabilistic semantics. The details are essentially a slight simplification of the development in Morgan (1991).

We are going to consider in a very general way (almost) arbitrary extensions to classical propositional logic. We wish to include in such extensions classical first-order predicate logic, higher order logics, as well as the usual quantified and non-quantified indexical logics.

Typically modal logics are given in axiomatic form and frequently include the rule of necessitation, which is validity preserving but not truth preserving. As a consequence, when dealing with indexical logics we usually think only of proofs and not of derivations from assumptions. But there is no need to be so restricted here, so in our development we will include considerations analogous to derivations from background assumptions.

From now on, we will assume that our language contains at least the syntactic machinery of classical sentence logic, including as primitive or defined classical conjunction ($\wedge$), disjunction ($\vee$), material implication ($\supset$), and negation ($\sim$). The language may be enriched with any other classical or non-classical connectives or quantifiers desired. We take the notion of derivation to be defined in the usual way as a finite sequence of WFFs, each member of the sequence being justifiable by appeal to an inference rule and previous members of the sequence. We use '$\vdash A$' to mean that A is a theorem. We assume that the proof theory can be specified by inference rules of the following sort:

IR. If $\vdash A_1$ and . . . and $\vdash A_j$, and A and the A_i satisfy conditions COND, then $\vdash A$.

Note that we will treat axioms as just special inference rules with vacuous antecedents. The designator COND in the above pattern is used to stand for any English sentence specifying well-founded special conditions (not involving semantic or proof-theoretic concepts) which must be satisfied. As an example, we can state a version of the standard rule for universal quantifier introduction as follows:

IR.UQI If $\vdash B \supset A$, and there are no free occurrences of variable x in B, then $\vdash B \supset (\forall x)A$.

The inference rules are stated in the metalanguage, and the A_i may be metalinguistic constants or variables. In the example IR.UQI above, the 'x', 'B' and 'A' are metalinguistic variables. Metalinguistic variables in the statements of the rules are assumed to be universally quantified.

Recall that we are considering extensions to classical propositional logic. At least part of what we mean is that every substitution instance of every classical propositional tautology is a theorem. We further assume that the usual inference patterns involving the classical sentential connectives are all admissible. However, we do not assume that the language must be restricted to, nor need it necessarily contain, sentential constants and variables.

If Γ is a finite non-empty set of WFFs, we use the notation $\wedge(\Gamma)$ to stand for the conjunction of the members of Γ. Since we are looking only at extensions of classical logic, the order and grouping of the conjuncts is logically irrelevant. If Γ is empty, then we may take $\wedge(\Gamma)$ to be some arbitrary tautology. For any set Γ, (empty, finite, or infinite) and WFF A, we use the notation $\Gamma \Rightarrow A$ to mean that there is some finite subset Δ of Γ such that $\vdash \wedge(\Delta) \supset A$.

From purely classical considerations, we know that our proof theory defines '$\Rightarrow$' as a standard syntactic entailment relation. In particular, the following lemmas are easily proved:

L.1 If $A \in \Gamma$, then $\Gamma \Rightarrow A$.

L.2 If $\Gamma \Rightarrow A$, then for some finite subset Δ of Γ, $\Delta \Rightarrow A$.

L.3 If $\Gamma \cup \{A\} \Rightarrow B$ and $\Gamma \Rightarrow A$, then $\Gamma \Rightarrow B$.

L.4 If $\Gamma \Rightarrow A$, then $\Gamma \cup \Delta \Rightarrow A$.

We say that a set of WFFs Γ is *consistent* iff there is at least one WFF, A, such that not $\Gamma \Rightarrow A$. We say that Γ is *maximally consistent* iff both of the following are satisfied: (i) Γ is consistent; and (ii) for every WFF B, if $\Gamma \cup \{B\}$ is consistent, then $B \in \Gamma$. We say that Γ is *maximal with respect to A* iff $A \notin \Gamma$ but Γ is maximally consistent.

Of course all of the above proof theoretic notation and definitions are relative to the specific logic under consideration. So that as we change the set of additional inference rules, we change what is to count as a proof, we change the pairs Γ and A for with $\Gamma \Rightarrow A$ holds, we change which sets are consistent and which sets are maximally consistent. Since we wish to remain perfectly general, we have not bothered to relativize our notation for reasons of simplicity.

Given all of our assumptions and definitions, the following theorems are easily established using L.1-4 and standard techniques:

L.5 If Γ is any consistent set of WFFs, then there is at least one maximally consistent extension of Γ.

L.6 If not $\Gamma \Rightarrow A$, then there is a superset Δ of Γ, such that Δ is maximal with respect to A.

L.7 $\Gamma \Rightarrow A$ iff for every maximally consistent superset Δ of Γ, $A \in \Delta$.

We will now turn our attention to a brief development of probabilistic semantics. In (Kolmogoroff 1950), Kolmogoroff presented his very influential, elegant formulation of *a priori* probability functions defined on a

σ-field of sets, along with the usual definition for conditional probability based on the *a priori* functions. We list the conditions here. For our purposes, it is sufficient to think of a σ-field of sets as any collection of arbitrary sets containing the empty set $\emptyset$, the universal set U, and closed under finitary unions and intersections.

KP.1 P is defined on a σ-field of sets.

KP.2 $0 \leq P(\alpha)$

KP.3 $P(U) = 1$

KP.4 If $\alpha \cap \beta = \emptyset$, then $P(\alpha \cup \beta) = P(\alpha) + P(\beta)$.

DKP.1 If $P(\beta) \neq 0$, then $P(\alpha, \beta) = P(\alpha \cap \beta)/P(\beta)$.

However, well known logical and mathematical problems arise from definition DKP.1 when the value of the conditioning set is 0. A very small change in Kolmogoroff's original formulation allows these problems to be completely avoided, without changing the underlying theory in any essential way. The essential idea for the appropriate change is due to Popper (1965), but this alternate formulation of the Kolmogoroff conditions is our own.

PP.1 P is defined on ordered pairs from a σ-field of sets.

PP.2 $0 \leq P(\alpha, \beta) \leq 1$

PP.3 $P(U, \alpha) = 1$

PP.4 If $\alpha \cap \beta = \emptyset$, then $P(\alpha \cup \beta, \gamma) = P(\alpha, \gamma) + P(\beta, \gamma)$, unless for all δ, $P(\delta, \gamma) = 1$.

PP.5 $P(\alpha \cap \beta, \gamma) = P(\alpha, \gamma) \times P(\beta, \alpha \cap \gamma)$.

If we then take the σ-field of sets to be the power set of the maximally consistent sets of formulas of classical logic, we can exploit the close connection between the classical sentence connectives and the usual operations on sets. We use superscript 'c' for set theoretic complement, and we use '$M(\gamma)$' for the set of maximally consistent supersets of Γ. The connections of interest can be stated as follows:

B.1 $M(\{A\}) \cup M(\{B\}) = M(\{A \vee B\})$

B.2 $M(\{A\}) \cap M(\{B\}) = M(\{A \wedge B\})$

B.3 $M(\Gamma) \cap M(\Delta) = M(\Gamma \cup \Delta)$

B.4 $M(\{A\})^c = M(\{\sim A\})$

B.5 If $\Gamma \subseteq \Delta$, then $M(\Delta) \subseteq M(\Gamma)$.

Using these relationships, we can derive a set of constraints for conditional probability theory over sentences of the logic, rather than over the σ-field of sets. It is this theory that we will use for our probabilistic semantics. The details of this entire development of our probability theory may be found in Morgan (1991).

The probability functions we will consider are defined on $\mathcal{L} \times \mathcal{P}(\mathcal{L})$. As indicated earlier, we read '$P(A, \Gamma) = r$' as 'the probability of A, given the assumptions in Γ, is r'. We may think of such functions intuitively as specifying the degree of rational belief one should have in conclusion A, given the premises Γ. It is not the task of logic to specify in general which sentences are actually true of our world; similarly it is not the task of probability theory to pick out one unique, preferred probability function. Rather the theory specifies rationality constraints which indicate relationships which must hold among the values assigned by any particular function in order for it to be considered rational.

We need to make one definition before stating the constraints, but first we will give a brief bit of background for it. Students of elementary logic (and indeed, some allegedly sophisticated philosophers) are frequently confused when they encounter the claim that from a logical falsehood, anything at all follows. That is, an argument with logically unsatisfiable premises and any conclusion whatever is automatically valid. One has to constantly remind such students that in order to be good, an argument must be reasonable not only in terms of the logical relation between premises and conclusion, but also in terms of the factual reasonableness of the premises; logically unsatisfiable premises are never factually reasonable. To put the matter more intuitively, if I cannot imagine any universe design in which the premises would all be true, then the argument should have no force in establishing that the conclusion is actually true of my own universe, and hence the fact that the argument is logically air tight is of no real consequence.

We can expand on these considerations a little. In judging evidential relationships, we must rely on some usually unstated background assumptions that are constitutive of the very concept of evidential relationship; such background assumptions may involve more than just logical principles. For example, I may doubt that my partner is being faithful and go about collecting evidence concerning this claim. But if at the same time I doubt the permanence of physical objects, it is not at all clear how I could gather evidence concerning the faithfulness of my partner. And if I doubt the principles of elementary arithmetic, it is very unclear how

I could judge any evidential relationships at all. Colloquially, we might say that if you would believe that an ordinary chair might at any moment turn into a charging lion, or that $1 + 1 \neq 2$, then you would believe anything. A set of premises so bizarre that they sanction any conclusion whatsoever is said to be an abnormal set. More precisely, we say that a set Γ is abnormal relative to probability function P (*P-abnormal*) iff for all sentences A, $P(A, \Gamma) = 1$. A sentence that is not P-abnormal is said to be P-normal. Where the specific probability distribution is clear, we use the simpler terminology 'normal' and 'abnormal'.

We are now in a position to state the details of our probability theory. For simplicity in stating the theory, we assume the language has definitions for all of the classical connectives in terms of $\sim$, $\wedge$, and $\vee$, and we will take these three as primitive. The following constraints define what we have called neo-classical conditional probability theory:

NP.1 $0 \leq P(A, \Gamma) \leq 1$.

NP.2 If $A \in \Gamma$, then $P(A, \Gamma) = 1$.

NP.3 $P(A \vee B, \Gamma) = P(A, \Gamma) + P(B, \Gamma) - P(A \wedge B, \Gamma)$.

NP.4 $P(A \wedge B, \Gamma) = P(A, \Gamma) \times P(B, \Gamma \cup \{A\})$.

NP.5 $P(\sim A, \Gamma) = 1 - P(A, \Gamma)$ provided Γ is P-normal.

NP.6 $P(A \wedge B, \Gamma) = P(B \wedge A, \Gamma)$.

NP.7 $P(C, \Gamma \cup \{A \wedge B\}) = P(C, \Gamma \cup \{A, B\})$.

These constraints by themselves turn out to be characteristic for classical propositional logic. If we define '$A \supset B$' as '$\sim A \vee B$', then it is easy to prove the following useful result using only NP.2-5:

L.8 If $P(A \supset B, \Delta) = 1$ for all Δ, then $P(B, \Delta \cup \{A\}) = 1$ for all Δ.

In order to obtain the theory appropriate for extensions to classical logic, we must also include for each additional inference rule of the form IR, above, a constraint of the following sort:

NP.IR If for all sets Δ, $P(A_1, \Delta) = 1$ and . . . and $P(A_j, \Delta) = 1$, and A and the A_i satisfy conditions COND, then for all sets Δ, $P(A, \Delta) = 1$.

Now we can turn our attention to the definition of semantic implication. First we will give a very brief motivation by considering a simple example. Usually when I flip a coin, I make the assumption that the coin will come up either heads (H) or tails (T). That is, if we let Γ be the description of the coin and the circumstances of the flip, I would assign $H \vee T$ the value 1, given Γ. If really pressed about the possibility of an alternative, I might say that while logically possible, I do not think there is any measurable probability that the coin will come up on edge (E); that is, I assign $\sim E$ the value 1, given Γ. Back to the wall, knife at my throat, I would have to admit that Γ does not logically entail $\sim E$ nor does it entail $H \vee T$. However, Γ does entail $H \vee \sim H$. The point is, there is information I could add to Γ that would make me doubt $H \vee T$; in particular, I would assign a probability of 0 to $H \vee T$, given $\Gamma \cup \{E\}$. But there is no evidence whatever that could be added to Γ that would make me doubt $H \vee \sim H$; that is, no matter what Δ contains, I would assign a value of 1 to the probability of $H \vee \sim H$, given $\Gamma \cup \Delta$. (A complete numerical treatment of this example may be found in (Morgan 1991).) So we should say that A is semantically implied by Γ only if there is no additional evidence we could add to Γ that would make us doubt A. Hence we adopt the following definition:

D.SI We say that the set of WFFs Γ *semantically implies* the WFF A iff for all probability functions P and all sets of WFFs Δ, $P(A, \Gamma \cup \Delta) = 1$. We use the notation '$\Gamma \Vdash A$' as shorthand for '$\Gamma$ semantically implies A'.

Of course our definition of semantic implication is relative to our probability theory, i.e., to the set of constraints. As we change the constraints, we change the allowed probability functions, and hence D.SI generalizes over a different set of functions as we vary the constraints. Since we are primarily interested in the general case, for simplicity we have not bothered to subscript our notation to indicate the dependence.

Note that in probabilistic semantics, it is appropriate to think of each probability function as analogous to a model. Each function describes one possible stochastic universe. Varying the background information is analogous to varying the world index in a possible worlds model. However, there is nothing in probabilistic semantics that corresponds to the accessibility relation between worlds. As we discussed in our motivation section, we may use the analogy to justify our definition of a canonical probability function.

D.SC We say that a probability distribution P is *strongly canonical* iff for every set of WFFs Γ and every WFF A: if not $\Gamma \Rightarrow A$, then there is some set of WFFs such that $P(A, \Gamma \cup \Delta) \neq 1$.

In other words, a strongly canonical distribution is one such that no matter what background assumptions we make, for any non-consequence of those assumptions, there is always some additional evidence we could find which should lead us to doubt that non-consequence. Thus strongly canonical distributions are minimalist from a logical point of view; they seem to completely commit us only to the actual logical consequences of our background assumptions.

3. Results

Our primary goal in this section is to prove completeness by proving the existence of strongly canonical distributions. However, we will first state and briefly discuss the soundness result.

Theorem 1. (soundness) If $\Gamma \Rightarrow A$, then $\Gamma \Vdash A$.

There are a number of possible approaches to proving soundness. Our conditions NP.1-7 are little different from those of (Popper 1965), and there are various soundness proofs in the literature relating classical propositional logic to Popper's conditions. Most any of these proofs can be easily modified for our constraints. Perhaps the easiest approach is just to adopt some standard axiomatic development of classical sentence logic formulated in terms of $\wedge$, $\vee$, and $\sim$, along with appropriate definitions for the other classical sentential connectives desired. A simple inductive argument on proofs using the results for classical logic plus NP.IR will then establish:

L.9 If A is an instance of a theorem of the logic, then for all probability functions satisfying NP.1-7 and NP.IR, $P(A, \Delta) = 1$ for all sets of WFFs Δ.

For the soundness theorem (Theorem 1, above), we consider an arbitrary set Γ and WFF A and suppose $\Gamma \Rightarrow A$. From the definition of '$\Rightarrow$' we know there is a finite subset Γ' of Γ such that $\vee(\Gamma') \supset A$ is a theorem. Then L.9 guarantees that for all sets of WFFs Σ, $P(\vee(\Gamma') \supset A, \Sigma) = 1$. And from this result, L.8 and NP.7 ensure that for all Σ, $P(A, \Sigma \cup \Gamma') = 1$. We obtain the desired result by considering sets Σ of the form $\Delta \cup \Gamma$.

We now return to a consideration of a proof of strong completeness by way of strongly canonical probability distributions. We begin by stating the strong completeness theorem.

Theorem 2. (strong completeness) If $\Gamma \Vdash A$, then $\Gamma \Rightarrow A$.

Following the standard Henkin argument, we seek to establish the contrapositive. We consider an arbitrary Γ and A such that not $\Gamma \Rightarrow A$, and we try to find some appropriate function P, satisfying NP.1-7 and the NP.IR such that $P(A, \Delta \cup \Gamma) \neq 1$ for some set Δ. This result would be established if we could prove the existence of a strongly canonical probability distribution. We will now prove the existence of such a distribution.

Theorem 3. There is at least one strongly canonical probability distribution satisfying NP.1-7 plus any collection of constraints of the form NP.IR.

Note that our probability distributions are not partial functions; they are defined everywhere. That means that even if our set of premises Γ is very limited in scope, we still have to be able to specify some degree of rational belief for every WFF in the language, even to those WFFs totally unrelated to our premises. And we cannot just arbitrarily assign all 0s or all 1s, since NP.5 requires that the values assigned to a formula and the negation of the formula must be complementary. In short, our problem is similar to the problem of making a complete truth value assignment when we start with only a small set of premises. And the solution in both cases is to consider maximally consistent extensions of the original premise set as the basis for making the required assignments. It is useful to consider two extreme suggestions for the use of maximally consistent sets in obtaining probability distributions.

The first suggestion is a technique which occurs in the literature in completeness proofs in cases in which the probability functions assign values to pairs of expressions, i.e., when the evidence is expressed in terms of a single WFF rather than as a set of WFFs. Given a maximally consistent set Γ, the proposal is to make the following assignment:

$$P(A, B) = \left\{ \begin{array}{l} 1 \text{ iff } B \supset A \in \Gamma \\ 0 \text{ otherwise.} \end{array} \right.$$

Quite apart from the problem of WFFs vs sets in the evidence position, this approach will not in general lead to a canonical distribution. Consider the case of classical logic, and suppose we take some arbitrary tautology as our evidence statement. Although every maximally consistent set is maximal with respect to *some* non-theorem, there is no maximally consistent set which is maximal with respect to *every* non-theorem of classical logic. So if we use only one maximally consistent set as the basis for our probability distribution, there will always be non-theorems which are assigned probability 1 on all evidence. Thus if we want to construct a canonical distribution, we must use more than one maximally consistent set as a basis.

The second suggestion is to take our lead from the canonical model construction technique for modal logics and to use all of the maximally consistent sets. Certainly if Γ is maximally consistent, then we could just make the following assignment:

$$P(A, \Gamma) = \left\{ \begin{array}{l} 1 \text{ iff } A \in \Gamma \\ 0 \text{ otherwise.} \end{array} \right.$$

But the problem is how to assign values when the evidence set is not maximally consistent. Since the evidence set may not even be deductively closed, much less negation complete, we cannot just extend this prescription to arbitrary evidence sets. An improved suggestion is to associate with each consistent set Γ some arbitrary maximally consistent extension $ME(\Gamma)$; if Γ is not consistent, we can just take $ME(\Gamma)$ to be the universal set of all WFFs. We could then make the following assignment:

$$P(A, \Gamma) = \left\{ \begin{array}{l} 1 \text{ iff } A \in ME(\Gamma) \\ 0 \text{ otherwise.} \end{array} \right.$$

This suggestion comes closer to working than our previous one, but is still not quite correct. Such a function would be canonical, but would not satisfy all of our constraints. The problem is best illustrated by considering NP.4 which we repeat here:

NP.4 $P(A \wedge B, \Gamma) = P(A, \Gamma) \times P(B, \Gamma \cup \{A\})$.

Suppose the left side has the value 1. Then $A \wedge B \in ME(\Gamma)$. It certainly follows that $A \in ME(\Gamma)$, so $P(A, \Gamma) = 1$. But there need be no relation whatever between $ME(\Gamma)$ and $ME(\Gamma \cup \{A\})$; and while A, B, and $A \wedge B$ may all be members of $ME(\Gamma)$, none of those facts imposes any requirement that B must be a member of $ME(\Gamma \cup \{A\})$. Simply note that Γ, A, and B may all be logically independent of each other. So our association of maximally consistent sets with evidence sets cannot be totally arbitrary if we wish to satisfy our constraints.

These considerations finally lead to a solution. Rather than considering arbitrary maximally consistent extensions of evidence sets, we will associate a preferred maximally consistent extension with each evidence set. This association will be done in such a way that as the evidence set is changed, the change in the associated extension will be minimal in a certain sense. There are a number of ways such a scheme could be implemented, but perhaps the simplest is based on the usual proof of the existence of maximally consistent extensions.

Suppose we are given some consistent set Γ, and we wish to prove that there must be a maximally consistent extension of Γ. Since the expressions of our language are countable, we consider some fixed enumeration of them; e.g., we could arrange them in some lexical order:

$$E_1, E_2, \ldots, E_j, \ldots$$

Although the list will not in general be finite, we know that for any expression in the language, that expression shows up on the list at some finite point. To "construct" a maximally consistent extension of Γ, we define a nested chain of sets as follows:

$$\Delta_0 = \Gamma$$
$$\Delta_i = \begin{cases} \Delta_{i-1} \cup \{E_i\} & \text{iff } \Delta_{i-1} \cup \{E_i\} \text{ is consistent} \\ \Delta_{i-1} & \text{otherwise.} \end{cases}$$

To obtain a maximally consistent extension of Γ, we just take the infinite union of all of the Δ_i; the proof that the union is indeed maximally consistent is the standard one and need not be repeated here. We may think of the enumeration of the expressions of the language as indicating some preference or importance ordering. In other words, we may think of the ordering as specifying a default ordering for the construction of model universes.

So, let us assume that the ordering of the E_i is fixed, indicating some default ordering. Then for every consistent set Γ, we will use the notation $DME(\Gamma)$ to stand for the unique maximally consistent extension of Γ which would be obtained by the indicated construction process, i.e., by adding to Γ in order each E_i that can be consistently added. For inconsistent sets Γ, we will take $DME(\Gamma)$ to be the set of all expressions $\mathcal{L}$. Intuitively, we think of $DME(\Gamma)$ as the default maximal extension of Γ. This technique of associating some default maximal extension with each set has the following desirable property:

L.10 If $A \in DME(\Gamma)$, then $DME(\Gamma) = DME(\Gamma \cup \{A\})$.

This theorem is easily proved by contradiction. Suppose $A \in DME(\Gamma)$, but $DME(\Gamma) \neq DME(\Gamma \cup \{A\})$. First note that our supposition guarantees that Γ is inconsistent iff $\Gamma \cup \{A\}$ is inconsistent. If Γ is inconsistent, then so will be $\Gamma \cup \{A\}$ by definition. On the other hand, since $A \in DME(\Gamma)$, it follows that if $\Gamma \cup \{A\}$ is inconsistent, so will be $DME(\Gamma)$ and thus Γ must also be inconsistent. Hence Γ is inconsistent iff $\Gamma \cup \{A\}$ is inconsistent. But if both are inconsistent, then their default maximal extensions are the same, contrary to supposition. Hence Γ and $\Gamma \cup \{A\}$ must both be consistent. Let Δ_i represent the sequence of nested sets in the construction of $DME(\Gamma)$, and let Σ_i represent the sequence of nested sets in the construction of $DME(\Gamma \cup \{A\})$. There must be some first expression E_k on which Δ_k and Σ_k differ. Then $\Delta_{k-1} = \Sigma_{k-1} - \{A\}$ or else Δ_{k-1} and Σ_{k-1} would be identical and could not differ with respect to consistency on the addition of E_k. Hence it must be the case that $\Delta_{k-1} \cup \{A\} \cup \{E_k\}$ is inconsistent, or else Δ_k and Σ_k would not differ with respect to E_k. So $E_k \in \Delta_k$. But then since $A \in DME(\Gamma)$, and $\Delta_k \subseteq DME(\Gamma)$, it follows that $DME(\Gamma)$ must be inconsistent, which is a contradiction. Thus our theorem is proved.

We can now return to a proof of Theorem 3, the existence of a strongly canonical probability distribution. We will use the default maximal extensions to define a probability distribution as follows:

$$P^*(A, \Gamma) = \left\{ \begin{array}{l} 1 \text{ iff } A \in DME(\Gamma) \\ 0 \text{ otherwise.} \end{array} \right.$$

Our first task is to prove that P^* satisfies the constraints NP.1-7 and all appropriate constraints of the form NP.IR. By definition, the values of P^* are limited to 1 and 0, so NP.1 is satisfied. That NP.2 is satisfied trivially follows from L.1 and L.7. The proofs for NP.3 and NP.5 depend on the usual principles concerning disjunctions and negations in maximally consistent sets. The proofs for NP.4 and NP.7 depend on L.10, as discussed above, and obvious principles of classical logic concerning conjunction. The proof that the NP.IR are satisfied depends on L.7 and the corresponding rules NP.IR. Thus P^* qualifies as an appropriate probability distribution for the logic. Our second task is to prove that P^* is strongly canonical. Consider an arbitrary WFF B and set of WFFs Δ such that not $\Delta \Rightarrow B$. By classical principles, we know that $\Delta \cup \{\sim B\}$ is consistent. Hence $B \notin DME(\Delta \cup \{\sim B\})$, and thus $P^*(B, \Delta \cup \{\sim B\}) = 0$. Thus P^* is strongly canonical. That completes the proof of Theorem 3.

Thus as promised, we have proved strong completeness by proving the existence of strongly canonical probability functions. The class of logics for which these results hold is extremely general, and includes modal logics which are not canonical with the usual possible worlds semantics.

One potentially interesting spin-off of this approach has been the characterization of default maximally consistent extensions. Such an approach might prove useful in the treatment of counterfactuals and of general default logics.

REFERENCES

Fine, K. (1975). Some Connections between Elementary and Modal Logic. In S. Kanger, ed., *Proceedings of the Third Scandinavian Logic Symposium*. Amsterdam: North Holland. pp. 15–39.

Hughes, G.E. and M.J. Cresswell (1968). *An Introduction to Modal Logic*. London: Methuen.

Kolmogoroff, A. (1950). *Foundations of the Theory of Probability*. New York: Chelsea.

Lemmon, E.J. and D. Scott (1966). *Intensional Logic*. Reprinted in Segerberg, K. (Ed.). (1977). *An Introduction to Modal Logic: The Lemmon Notes*. Oxford: Basil Blackwell.

Morgan, C.G. (1982). There Is a Probabilistic Pemantics for Every Extension of Classical Sentence Logic. *Journal of Philosophical Logic* **11**, 431–442.

Morgan, C.G. (1991). Logic, Probability Theory, and Artificial Intelligence — Part I: The Probabilistic Foundations of Logic. *Computational Intelligence* **7**, 94–109.

Popper, K. (1965). *The Logic of Scientific Discovery*. New York: Harper and Row.

Segerberg, K. (1968). Decidability of S4.1. *Theoria* **34**, 7–20.

Segerberg, K. (1971). *An Essay in Classical Modal Logic*. Uppsala.

Charles G. Morgan
Department of Philosophy
University of Victoria
P.O. Box 3045
Victoria, B.C. Canada V8W 3P4
and
Varney Bay Institute for Advanced Study
P.O. Box 45
Coal Harbour, B.C. Canada V0N 1K0
e-mail: morgan@uvphys.phys.uvic.ca

Commentary by François Lepage

Morgan's proposal is a piece of a vast enterprise: the characterization of the concept of validity using the notion of probability without any reference to the notions of truth and falsity.

In preceding papers, Morgan showed that for (almost) all extensions of classical logic, there is a probabilistic semantics which is characteristic for that logic. Here, Morgan goes one step further: for each of these logics, there is a *canonical probability distribution*, that is a probability distribution P such that for each non-theorem A, there is a set of sentences Γ such that $P(A, \Gamma) \neq 1$. This means that for any non theorem A there is a set of evidence that provides grounds for doubting A. Morgan shows, in fact, a little bit more. For each of these logics, there is a *strongly canonical probability distribution*, that is, a probability distribution P such that for each A and any set of sentences Γ such that it is not the case that $\Gamma \Rightarrow A$, there is a set of sentences Δ such that $P(A, \Gamma \cup \Delta) \neq 1$. This means that there are always sets of evidences that provide grounds for doubting anything excepted the logical consequences of what we presuppose.

Morgan's result is quite interesting in itself, in that it provides a characterization of a wide class of logics without using any notion of truth. Beside the technical questions, Morgan's approach is a step toward a kind of naturalization of the central concept of logic in term of belief, a step that the classical semantic approach cannot make. Let's see why.

From a naturalistic point of view, logic should be grounded in one way or another on (an idealization of) the rational behavior of agents. If we take the semantic approach seriously, together with the notion of a model or interpretation where the value of a sentence is ultimately a truth value (or a truth value in each accessible world, etc.), we rapidly meet the problem of logical omniscience and of omniscience *tout court*. A valid sentence is a sentence which is true in any interpretation. But, what is a valid sentence from a naturalistic point of view? Probably a sentence that an agent must take to be true (I don't know exactly what modality is expressed here by 'must') or a sentence a rational agent should believe. This raises a very hard problem. Let p be any sentence. According to classical logic, $p \lor \neg p$ is a valid sentence. But this is not compatible with the classical semantic approach because according to this view, any sentence of the form $q \lor r$ is true if, and only if, q is true or r is true. This means that, $p \lor \neg p$ is true if, and only if, p is true or $\neg p$ is true (i.e. p is false). So, from a naturalistic point of view, a rational agent must believe p or $\neg p$. We have to choose between two alternatives: (1) to give up classical logic or (2) to accept omniscience. There is naturally a third option, to give up semantics. This is, by the way, the approach chosen in the AGM modeling of belief where the epistemic states are sets of sentences which are deductively closed on a logic which contains classical logic. These epistemic states *cannot* be models except in case they are maximally consistent sets, and this is omniscience.

Let's go back to Morgan. Is his approach a better tool for a naturalist? If we take a strongly canonical probability distribution as a candidate for a belief function of an epistemic agent, the situation is even worse than in the classical semantic approach. An agent is not omniscient but an omniconditionalizer! But Morgan is proposal authorizes another interpretation. Let's suppose that an agent's probability function satisfy NP.1 to NP.2 but is partial. In other words, the agent's function satisfies NP.1 to NP.2 when it is defined, but it is not always defined. We add the constraint that $P(A, \Gamma)$ is defined if, and only if $P(B, \Gamma')$ is defined, where $B \in \Gamma \cup A$ and $\Gamma' \subseteq \Gamma \cup A$ and Γ' is not abnormal. We can similarly define the notion of semantic implication for partial probability function by adding to D.SI the clause 'when defined'. The domain D of sentences where P is defined can clearly be extended to the set of set of all sentences. Then Morgan's constructions ride again. Such a probability function P can always be extended to a strongly canonical probability function.

The moral of this story is the following. A rational agent does not have to be either omniscient nor logically omniscient. His probability function can be undefined even on tautologies. But a non valid sentence A is one such that there is an extension of the belief function of the agent, a set Γ of sentences (on which the agent may have no opinion) such that Γ provides some grounds for doubting of A. And this can be done in the framework of classical logic.

François Lepage
Department of Philosophy
University of Montreal
Montreal, P.Q. H3C 3J7 Canada
e-mail: lepagef@ere.umontreal.ca

Morgan's Reply to Lepage

Prof. Lepage is certainly correct when he points to the strong temptation to substitute "rational belief systems" for classical models when trying to develop a naturalistic characterization of central logical concepts. And he correctly identifies one of the major critiques of probability theory as an account of rational belief systems: namely, probability distributions are overly determined when compared to the beliefs of real agents.

One serious problem with classical model theory as the basis for an account of real reasoning is that it does seem to require logical omniscience. But as Lepage points out, numerical conditional probability theory requires even more knowledge! Simply note that every maximally consistent set of expressions (roughly a classical model) corresponds to an *a priori* probability distribution. So a conditional probability distribution requires the same amount (by some intuitive measure) of knowledge as that required for a classical model, plus that same amount of knowledge for *each* possible set of background assumptions. In the very descriptive phraseology of Lepage, probability theory requires one to be an omniconditionalizer.

Lepage suggests a partial solution to the problem of over specificity of probability distributions by allowing partial distributions. Lepage's criteria would ensure that a partial distribution could always be extended to a complete distribution. I am much in favor of the direction in which Lepage is asking us to move, and I think even more movement in that direction should be considered. There are two fronts on which advances should be made.

The first front is that considered by Lepage. His suggestion is (roughly) that assigning a degree of belief to every element $\Gamma \cup \{A\}$, given an arbitrary subset Γ', should be necessary and sufficient for assigning a degree of belief to A, given Γ. If we are trying to accord with real belief systems, I suspect that even this suggestion is too strong. Real humans may assign degrees of belief to every element of the set $\Gamma \cup \{A\}$, given some subset, without having *considered* the probability of A, given Γ. Essentially, Lepage's suggestion is that if a value *could* be computed, then is must be assigned, which still requires too much detail to be realistic. In short, I think we need to allow distributions to be even more partial than Lepage's condition would permit. Basically what is needed is a set of easily checked conditions that would allow building up a probability distribution one value at a time, in such a way that one is assured that as long as the conditions are satisfied, there is some way to extend the partial distribution to a total distribution. Alas, I do not have such a set of conditions; it remains an open area for research.

The second front is to abandon numerical values. Human beings cannot linearly order their beliefs, much less assign precise numerical weights. So a consideration of comparative probability theory would be more realistic than the usual numerically valued theories. We have made some progress on this second front; see "Comparative Probability as a Foundation for Logic," forthcoming. In the comparative theory, there is no requirement for pairwise comparability, so the comparative relation may be extremely sparse, or "partial." But even so, in the current account the comparative distributions are still too overly specified to be realistic, and conditions permitting building up a comparative relation one step at a time are needed, as Lepage would suggest.

Poznań Studies in the Philosophy of the Sciences and the Humanities
2000, vol. 71, pp. 36–48

François Lepage

A MANY-VALUED PROBABILISTIC
CONDITIONAL LOGIC

ABSTRACT. Following Lewis, Gärdenfors and Nute, we present a conditional logic in which the probability of a conditional is obtained by imaging, i.e., the probability of the conditional is the probability of B after distribution of the probability of the non-A-worlds on the nearest A-worlds. We show that this is possible in Lewis' System of Spheres Semantics only if one use a many-valued logic where the truth value of the conditional is the ratio of the number of B-worlds to the number of A-worlds in the smallest sphere containing A-worlds. A system is provided which is proved to be a conservative extension of the classical propositional logic.

1. Prolegomenas

In his celebrated paper, "Probability of Conditionals and Conditional Probabilities," David Lewis (1976) showed that, contrary to a claim of Robert Stalnaker's (1968; 1970), it is not possible to introduce into the classical propositional calculus a counterfactual conditional "if A were the case, B would be the case" (hereafter "$A > B$") in order to obtain a language and a semantics for that language having the following properties:

(1) the classical part behaves classically, i.e., any proposition which contains no counterfactual has its classical truth conditions;

(2) starting from any possible world, possible worlds are linearly ordered, so we can speak of the distance of a world from the starting world, and the truth conditions of the counterfactual are the following: $A > B$ is true in a world w iff either B is true at the closest world to w where A is true (hereafter "A-world") or there is no A-world;

(3) any probability function defined on the classical part can be extended to the set of all propositions by stipulating that the probability of the counterfactual is the probability of the consequent, given the antecedent i.e., $P(A > B) = P(B/A)$ if $P(A) \neq 0$.

Clause (2) will be called Stalnaker's clause. Lewis has shown that if clauses (1), (2) and (3) are not, strictly speaking, inconsistent, they can be satisfied only by trivial probability functions, i.e., by probability functions which give to any proposition either the value 0 or the value 1.

In the same paper, Lewis suggested another way to interpret the probability of a conditional, preserving clauses (1) and (2) and giving up clause (3).

Let P be a probability function defined on the classical part. Lewis suggested an extension of P in the following way:

(4) $P(A > B) = P_A(B)$ where P_A is the probability function obtained from P by shifting the probability of any $\neg A$-world on the nearest A-world according to Stalnaker's clause.

Lewis has shown that this definition is compatible with the general constraint that the probability of any proposition A is the sum of the probability of the A-worlds. This is called *Imaging*.

But Stalnaker's clause is far from being intuitive. If in some situations, there is clearly a nearest A-world, this does not obtain generally. Let's suppose a dice is thrown and gives a 6. It is quite natural to think that there is *one* closest world where the throw gave, say, 5.

What, now, about the closest world where the throw didn't give a 6? According to Stalnaker's clause, the worlds where the throw gave 1, 2, 3, 4 or 5 are linearly ordered: one of them is the nearest, another the second nearest, etc. For obvious reasons of symmetry, this consequence of Stalnaker's clause is unintuitive and casts discredit on the clause itself.

Curiously, Lewis (1973) had already provided a semantics for counterfactuals that can bypass this difficulty, namely his System Of Spheres semantics (hereafter SOS).

In this semantics, possible worlds are not linearly ordered, but weakly ordered, i.e., many worlds may be at the same distance from a given world. The best image is of embedded spheres of possible worlds centered on the world of evaluation. All the worlds of a given layer are equidistant from the world of evaluation. The truth conditions of the counterfactual are then the following:

(5) $A > B$ is true in a world w iff B is true at *all* the nearest A-worlds to w or there is no A-world.

Unfortunately, imaging is not compatible with Lewis' SOS semantics.

It can be shown (Lewis 1997) that if we generalize imaging as defined by clause (4) by sharing out the probability of any world on the nearest worlds where the antecedent is true, then the probability of a conditional is the probability of the consequent after this sharing out if and only if each layer of the system of spheres around any world contains exactly one world. In short: imaging works in a SOS if and only if this SOS is a Stalnaker system.

But there is another way to introduce imaging in SOS. It is by changing the truth conditions of the conditional. Let $A > B$ have as its truth value the ratio of $A \wedge B$-worlds on the A-worlds of the smallest sphere having at least one A-world. In the limit cases where all A-worlds are $A \wedge B$-worlds or no A-worlds are $A \wedge B$-worlds, the truth value of the conditional is the same as in Lewis' SOS, i.e., 1 or 0. In the intermediary cases where some but not all of the A-worlds are $A \wedge B$-worlds, the conditional will have a fractionary value. A brief presentation of that semantics will be the task of the first part of my paper.

The introduction of fractionary values raises the well-known problems associated with many-valued logic, that is, there is no extensional many-valued extension of the classical logic in the following sense:

(5) All instances of tautologies are valid;

(6) If all the sub-expressions of a proposition have classical truth values, then this proposition also has a classical truth value.

In the second part of the paper, I present a *non-extensional* logic satisfying (5) and (6) which uses a counterfactual for the definition of the truth conditions for conjunction.

2. Imaging in a SOS

Let's first present the syntax of the system. The set of atomic propositions is $A = \{p_i\}_{i \in n}$ where n is a finite number, and the set of propositions is the smallest set L such that

(i) for any $i \in n$, $p_i \in L$

(ii) if $A \in L$, then $\neg A \in L$

(iii) if $A, B \in L$, then $(A \wedge B), (A > B) \in L$.

We can now provide the system with an interpretation. Let W be the set of possible worlds defined as $W = \{0, 1\}^A$ (where 0 and 1 respectively express the falsity and the truth) and let $f : L \times W \to \wp(W)$. f is a *selection function* such that, for any world w and any proposition A, $f(A, w)$ is a set of worlds containing at least one A-world. I will consider two classes of selection functions. Firstly, I will consider *Stalnaker functions*, where f

selects, for any $\langle A, w \rangle$, one and only one world. All the worlds selected by varying A are linearly ordered (the closest world being the actual world).

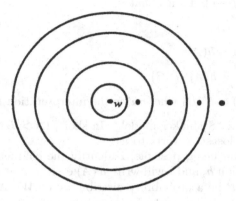

Figure 1.

For a Stalnaker function, $f(\langle A, w \rangle)$ is the closest A-world to w (or equivalently, the smallest set containing one A-world).

I will also consider *Lewis selection functions*. Selected worlds are in embedded spheres and the smallest sphere contains only the actual world. Thus a given Lewis selection function selects for each A the smallest sphere containing at least one A-world.

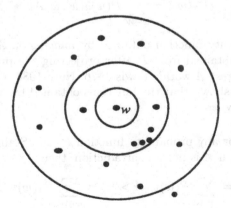

Figure 2.

Stalnaker functions are just a special case of Lewis functions, the case where each layer contains exactly one world.

Let f be a Stalnaker function. An interpretation based on $w \in W$ is a function $h : L \to 0, 1$ such that

(i) $h(p_j) = w(p_j)$

(ii) $h(\neg A) = 1 - h(A)$

(iii) $h(A \wedge B) = h(A) \cdot h(B)$

(iv) $h(A > B) = h'(B)$ where h' is the interpretation based on $f(A, w)$.

We will call h a "Stalnaker model." In short, $(A > B)$ is true at w iff B is true at the closest A-world to w.

No confusion being possible, I identify the characteristic function w with its extension h, and I will write $w(A)$ even in the case where A is not an atom. Let P be a probability distribution on W. As usual, we define $P(A) = \sum_{w \in W} P(w)w(A)$ and so we trivially have

(i) $P(A) = 1$ if A is a tautology

(ii) $P(\neg A) = 1 - P(A)$

(iii) $P(A \vee B) = P(A) + P(B)$ if $w(A \wedge B) = 0$.

Let δ be a function such that $\delta(w', w, A) = 1$ iff $w' = f(\langle A, w \rangle)$. Following Lewis (1976), I define

$$P_A(w') = \sum_{w \in W} P(w)\delta(w', w, A)$$

P_A is the probability function obtained by *imaging* on A, i.e., the probability function obtained from P after projecting the probability of any world on the nearest A-world (Lewis 1976; Nute 1980; Gärdenfors 1982, 1988). Lewis has shown that the functions obtained by imaging have the following property:

Proposition. For any probability function P, any Stalnaker function f and any world w, if A is not a contradiction, then

$$(*) \quad P(A > B) = \sum_{w \in W} P(w)w(A > B) = \sum_{w \in W} P_A(w)w(B) = P_A(B).$$

This technique is radically different from conditionalization as shown by Gärdenfors (1982).

The restriction to Stalnaker functions, i.e., the hypothesis that, from the point of view of any world, W is linearly ordered, is a very constraining one. It would be very interesting to obtain a similar result using Lewis functions. Unfortunately, as suggested by Nute (1980), it is not possible.

Let us define a Lewis model (Lewis 1973). It is a model similar to the one above except that f is a Lewis function and (iv) becomes

(iv') $w(A > B) = 1$ iff $w'(B) = 1$ for any $w' \in f(\langle A, w \rangle)$ such that $w'(A) = 1$.

For example, in the following situation we have $w(A > B) = 0$ and $w(B > A) = 1$.

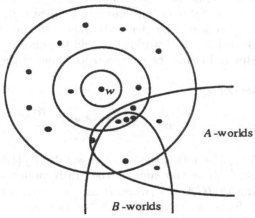

Figure 3.

We could try to adapt the imaging technique to Lewis models by defining

$$(**) \qquad P_A(w') = \sum_{w \in W} P(w)\delta(w', w, A)c_{w,w',A}$$

where the $c_{w,w',A}$ are weighting coefficients, i.e., for any w,

$$\sum_{w' \in W} \delta(w', w, A)c_{w,w',A} = 1.$$

Hence we obtain the following result (Lepage 1997):

Proposition. For any Lewis model and any P, the function P_A obtained by *imaging* according to $(**)$ satisfies the equation

$$P(A > B) = \sum_{w \in W} P(w)w(A > B) = \sum_{w \in W} P_A(w)w(B) = P_A(B)$$

iff f is a Stalnaker function.

The reason for this is simple: According to the truth conditions of the Lewis conditional, when the smallest sphere around w containing at least one A-world contains $(A \wedge B)$-worlds and $(A \wedge \neg B)$-worlds the conditional is false, so $P(w)w(A > B) = 0$ and the probability projected on A-worlds is lost. Thus, in that case $P(A > B) < P_A(B)$. Therefore, imaging is not compatible with Lewis' original semantics.

Fortunately, if we modify this semantics, imaging is possible again. The modification consists in allowing conditionals to take fractional truth values. The truth value of a conditional $(A > B)$ is the ratio of the number of $(A \wedge B)$-worlds on the number of A-worlds in the smallest sphere containing at least one A-world. When all the A-worlds are B-worlds or no A-worlds are B-worlds, we find the classical truth values 1 and 0 again. But in halfway cases, the truth value of the counterfactual is a fraction.

Formally, we define

$$w(A > B) = \frac{\sum_{w' \in V(A,w)} w'(B)}{n_{A,w}}$$

where $V(A, w) \subseteq f(\langle A, w \rangle)$ is the set of A-worlds of $f(\langle A, w \rangle)$.

$\sum_{w' \in V(A,w)} w'(B)$ is the sum of the truth values of B in the A-worlds belonging to $f(\langle A, w \rangle)$ (when B has not itself a fractional value, $\sum_{w' \in V(A,w)} w'(B)$ is just the number of A-worlds among the A-worlds of $f(\langle A, w \rangle)$; $n_{A,w}$ is the number of A-worlds in $f(\langle A, w \rangle)$).

Look again at Figure 3. Here, the smallest sphere containing at least one A-world contains four A-worlds, three of them are B-worlds and thus $w(A > B) = 3/4$.

With this new definition of $w(A > B)$, it is easily proved that $(*)$ holds again. But now, what happens to complex propositions? The natural truth conditions of negation are surely given by $w(\neg A) = 1 - w(A)$. There is no natural definition of truth conditions for the conjunction. This is a very well-known problem for any many-valued logic (Urquhart 1986). One can easily show that *no* extensional definition of conjunction results in a many-valued logic which is an extension of classical logic, i.e., one where

(i) any instance of a tautology is valid and two tautologically equivalent expressions are equivalent;

(ii) any formula in which only classical connectors have occurences takes its classical value for any classical valuation.

This brings us to the second part of our paper. I will now introduce the notion of *normal form* (hereafter *NF*) for propositions of *L* and provide an interpretation for these normal forms using the counterfactual truth conditions to define the truth conditions of the conjunction.

3. A New Semantics for Conjunction

An *NF* is just a slight modification of the notion of *full normal disjunctive form* in order to take into account the occurrences of counterfactuals. By a full normal disjunctive form of a *classical* formula *A*, I mean the following: If *A* is a contradiction, the full normal disjunctive form of *A* is **0** (a canonical name for falsity); the full normal disjunctive form of ¬**0** is **1** (a canonical name for truth); otherwise the full normal disjunctive form of A is the shortest formula which is tautologically equivalent to *A* of the form $\neg(\neg A_1 \wedge \cdots \wedge \neg A_m)$ where each A_j is a conjunction of literals in some canonical order.

In the general case, we define a function $NF : L \to L$ such that

(i) $NF(l_i) = l_i$ (where l_i is a literal, i.e., p_j or $\neg p_j$ for some j)

(ii) $NF(A > B) = NF(A) > NF(B)$

(iii) If *A* is not as in (i) and (ii), then $NF(A)$ is the full normal disjunctive form of *A* where any counterfactual $C > B$ is treated like an atom and is replaced by $NF(C > B)$.

So, any *NF* is of the following form: $\neg \bigwedge_{j=0}^{m-1} \neg A_j$ where A_j is itself a conjunction of literals or of counterfactuals in normal form. We can now define an interpretation for formulas in normal form. We will need the following tools.

An interpretation based on $w \in W$ is defined as usual, with the additional hypothesis that the SOS is such that if *A* and *B* have no atom in common, then the atoms of *B* have the same value in *w* and in the closest *A*-worlds to *w*. This constraint can be interpreted as meaning that if *A* and *B* are independent, then $w(A > B)$ is just $w(B)$. So,

(i) $w(\neg A) = 1 - w(A)$

(ii) $w(A > B) = \dfrac{\sum_{w' \in V(A,w)} w'(B)}{n_{A,w}}$ (as above)

(iii) Let $\bigwedge\limits_{j=0}^{m-1} A_j$ be a conjunction of m propositions. We define

$$w\left(\bigwedge_{j=0}^{m-1} A_j\right) = \sqrt[m]{\prod_{j=0}^{m-1} w(A_j) \cdot \prod_{j,k=0}^{m-1} w(A_j > A_k)}.$$

The idea is to interpret a conjunction of m formulas not only as the logical product of the values of the conjuncts but to take also into account a kind of "proximity" between the conjuncts, which is express by $w(A_j > A_k)$ [1].
Let us consider the following examples:

(1) $w(p_j \wedge p_i) = \sqrt{w(p_i) \cdot w(p_j) \cdot w(p_i > p_j) \cdot w(p_j > p_i)}$
    $= \sqrt{w(p_i) \cdot w(p_j) \cdot w(p_j) \cdot w(p_i)} = w(p_j) \cdot w(p_i)$

(2) $w(p_j \wedge p_j) = w(p_j)$

(3) $w(p_j \wedge \neg p_j) = \sqrt{w(p_j) \cdot w(\neg p_j) \cdot w(p_j > \neg p_j) \cdot w(\neg p_j > p_j)}$
    $= \sqrt{w(p_j) \cdot w(\neg p_j) \cdot 0 \cdot 0} = 0$

(4) $w(p_j \vee \neg p_j) = 1$

(5) $w(p_j \wedge p_i \wedge p_k) = \sqrt[3]{w(p_j) \cdot w(p_i) \cdot w(p_k) \cdot w(p_j > p_i) \cdot w(p_i \cdot p_j)} \cdot$
    $\overline{w(p_j > p_k) \cdot w(p_k > p_j) \cdot w(p_k > p_i) \cdot w(p_i > p_k)}$
    $= w(p_j) \cdot w(p_i) \cdot w(p_k)$

(6) In the general case, if A_k and A_i have no atom in common,

$$w\left(\bigwedge_{j=0}^{m-1} A_j\right) = \prod_{j=0}^{m-1} a(A_j).$$

Unfortunately, the definition of the truth conditions for the conjunction given above is not recursive, because it depends on the number m of conjuncts. This difficulty can easily be bypassed. Let us define the following two functions:

$Conj(A)$ is a function which counts the number of conjuncts in A.

$Comp(A)$ is the set of conjuncts of A.

$Conj(l) = 1$ (when l is a literal)

$Conj(A \wedge B) = Conj(A) + Conj(B)$

$Conj(\neg A) = 1$

$Conj(A > B) = 1$

$Comp(l) = \{l\}$

$Comp(\neg A) = \{\neg A\}$

$Comp(A \wedge B) = Comp(A) \cup Comp(B)$

---

[1] The idea to use counterfactuals as a measure of proximity was suggested to me by Professor Jian-Yun Nie.

$Comp(A > B) = \{(A > B)\}$

Using these two functions, we can replace the truth condition for the conjunction given in (iii) by

(iv) $w(A) = \ _{Conj(A)}\sqrt{\prod_{A_j \in Conj(A)} w(A_j) \cdot \prod_{A_j,k \in Comp(A)} w(A_j > A_k)}.$

(i), (ii) and (iv) are recursive clause that provide a value for any NF. We obtain the following results.

Proposition. Any instance of a tautology is valid.

This property is trivial since we work with NF and the NF of any instance of a tautology is 1.

The next result is (at first sight) less trivial:

Proposition. Any classical proposition (in which there is no occurrence of conditionals) has its classical truth conditions.

Proof. We just have to check that clauses (i) and (iv) behave classically for classical arguments, and this is straightforward.

This last result means that the complicated truth conditions given by (i), (ii) and (iv) are just the classical ones when all the arguments are classical ones. So, this semantics is an extension of the classical one.

Let us now turn to the question of providing a system for that logic. It is an interesting fact that for any axiom of the complete system of Stalnaker's logic, the corresponding rule is valid (by the corresponding rule I mean the replacement of $\vdash A \supset B$ by $A \vdash B$).

Proposition. All the rules of the (complete) Stalnaker system are valid, i.e.,

$\vdash \top$ if $\top$ is an instance of a tautology

$((A > B) \wedge (A > C)) \vdash (A > (B \wedge C))$

$\vdash A > \top$

$\vdash A > A$

$(A \wedge B) \vdash (A > B)$

$(A > B) \vdash (A \supset B)$

$((A > C) \wedge (B > C)) \vdash ((A \vee B) > C)$

$((A > B) \wedge (A > C)) \vdash ((A \wedge C) > B)$

$\vdash ((A > B) \vee (A > \neg B))$

If $\vdash (B \supset C)$, then $\vdash (A > B) \supset (A > C)$

If $\vdash (A \equiv B)$, then $\vdash (A > C) \supset (B > C)$

If $\vdash (A \supset B)$ and $\vdash A$, then $\vdash B$.

The law of the conditional excluded middle is valid because $w(A > B)$ and $w(A > \neg B)$ are always complementary numbers, i.e., they sum up to 1. A completeness proof is still to come.

REFERENCES

Gärdenfors, P. (1982). Imaging and Conditionalization. *The Journal of Philosophy* **79**, 747–760.

Gärdenfors, P. (1988). *Knowledge in Flux: Modeling the Dynamics of Epistemic States.* Cambridge: MIT Press.

Lepage, F. (1997). Conditionals, Imaging, and Subjunctive Probability. *Dialogue* **36**, 113–135.

Lewis, D. (1976). Probability of Conditionals and Conditional Probabilities. *Philosophical Review* **85**(3), 297–315.

Lewis, D. (1973). *Counterfactuals.* Cambridge: Harvard University Press.

Nute, D. (1980). *Topics in Conditional Logic.* Dordrecht: Reidel.

Stalnaker, R. (1968). A Theory of Conditionals. In N. Rescher (Ed.), *Studies in Logical Theory.* Oxford: Blackwell. pp. 98–112.

Stalnaker, R. (1970). Probability and Conditionals. *Philosophy of Science* **37**(1), 64–80.

Urquhart, A. (1986). Many-Valued Logic. In D. Gabbay and F. Guenthner (Eds.), *Handbook of Philosophical Logic, Vol. 3: Alternatives to Classical Logic.* Dordrecht: Reidel. pp. 71–116.

François Lepage
Department of Philosophy
University of Montreal
Montreal PQ H3C 3J7 Canada
e-mail: lepagef@ere.umontreal.ca

Commentary by Charles G. Morgan

Prof. Lepage is certainly correct when he points to the strong temptation to substitute "rational belief systems" for classical models when trying to develop a naturalistic characterization of central logical concepts. And he correctly identifies one of the major critiques of probability theory as an account of rational belief systems: namely, probability distributions are overly determined when compared to the beliefs of real agents.

One serious problem with classical model theory as the basis for an account of real reasoning is that it does seem to require logical omniscience. But as Lepage points out, numerical conditional probability theory requires even more knowledge! Simply note that every maximally consistent set of expressions (roughly a classical model) corresponds to an *a priori* probability distribution. So a conditional probability distribution requires the same amount (by some intuitive measure) of knowledge as that required for a classical model, plus that same amount of knowledge for *each* possible set of background assumptions. In the very descriptive phraseology of Lepage, probability theory requires one to be an omniconditionalizer.

Lepage suggests a partial solution to the problem of over specificity of probability distributions by allowing partial distributions. Lepage's criteria would ensure that a partial distribution could always be extended to a complete distribution. I am much in favor of the direction in which Lepage is asking us to move, and I think even more movement in that direction should be considered. There are two fronts on which advances should be made.

The first front is that considered by Lepage. His suggestion is (roughly) that assigning a degree of belief to every element $\Gamma \cup \{A\}$, given an arbitrary subset Γ', should be necessary and sufficient for assigning a degree of belief to A, given Γ. If we are trying to accord with real belief systems, I suspect that even this suggestion is too strong. Real humans may assign degrees of belief to every element of the set $\Gamma \cup \{A\}$, given some subset, without having *considered* the probability of A, given Γ. Essentially, Lepage's suggestion is that if a value *could* be computed, then ii must be assigned, which still requires too much detail to be realistic. In short, I think we need to allow distributions to be even more partial than Lepage's condition would permit. Basically what is needed is a set of easily checked conditions that would allow building up a probability distribution one value at a time, in such a way that one is assured that as long as the conditions are satisfied, there is some way to extend the partial distribution to a total distribution. Alas, I do not have such a set of conditions; it remains an open area for research.

The second front is to abandon numerical values. Human beings cannot linearly order their beliefs, much less assign precise numerical weights. So a consideration of comparative probability theory would be more realistic than the usual numerically valued theories. We have made some progress on this second front; see "Comparative Probability as a Foundation for Logic," forthcoming. In the comparative theory, there is no requirement for pairwise comparability, so the comparative relation may be extremely sparse, or "partial." But even so, in the current account the comparative distributions are still too overly specified to be realistic, and conditions permitting building up a comparative relation one step at a time are needed, as Lepage would suggest.

Charles G. Morgan
Department of Philosophy
University of Victoria
P.O. Box 3045
Victoria, B.C. Canada V8W 3P4
and
Varney Bay Institute for Advanced Study
P.O. Box 45
Coal Harbour, B.C. Canada V0N 1K0
email: morgan@uvphys.phys.uvic.ca

Lepage's Reply to Morgan

I agree with Prof. Morgan that there is no obligation to consider fractionary values as truth-values. It is in fact possible to deny either that counterfactuals have truth values or have Levis' truth values, while still using imaging to calculate the probability of the counterfactual. The advantage of considering fractions of truth values is that it gives us a straightforward generalization of Lewis' imaging technique, where the truth value itself is used as a characteristic function.

But there is also an advantage, as suggested by Prof. Morgan, in not considering fractional truth values. In fact, the only constraint for imaging in a SOS, is that no probability should be lost or created. For example, it is possible to share the probability of $\neg A$-world on the A-worlds of the smallest sphere proportionally to their old probability. Such an approach will remove the difficulty with the conjunction.

*Poznań Studies in the Philosophy
of the Sciences and the Humanities*
2000, vol. 71, pp. 49-76

Piers Rawling

THE EXCHANGE PARADOX, FINITE ADDITIVITY, AND
THE PRINCIPLE OF DOMINANCE [1]

ABSTRACT. This essay concerns the exchange paradox. Part of my purpose here is to provide a review of previous results. In addition, I hope to clear up what I see as some misconceptions concerning the problem. There are two different problems lurking here. The first is easily dismissed; the second raises deeper issues. Within a countably additive probability framework (see Section 3), the second problem illustrates a breakdown of dominance with respect to infinite partitions in circumstances of infinite expected utility. Within a probability framework that is only finitely additive (see Section 4), there are failures of dominance with respect to infinite partitions in circumstances of bounded utility with finitely many consequences.

1. Introduction

Ten dollars have been placed in an envelope (call it 'O' for 'original'). And a fair coin has been tossed. If it came up heads, twenty dollars were placed in a second envelope (call it 'T' for 'toss'). If it came up tails, five dollars were placed in T. You are given one of the envelopes; you know not which. You are then given the opportunity to trade your envelope for the other of the pair. You reason as follows. There is a 50% chance that T contains half the contents of O, and a 50% chance that T contains twice the contents of O. Thus the expected actuarial value ("*eav*") of T is 1.25 times $((0.5 \times 2) + (0.5 \times 0.5))$ that of O. But there is also a 50% chance that O contains half the contents of T, and a 50%

---

[1] I have discussed the two envelopes problem with many people. Particular thanks are owed to John Broome, Jamie Dreier, Kirk Ludwig, Ned McClennen, Al Mele, Greg Ray, Teddy Seidenfeld, Susan Vineberg, and audiences at the 1996 meeting of the Society for Exact Philosophy and the December 1996 meeting of the British Society for the Philosophy of Science. My source for Ian Hacking's contribution is a letter summarizing his example that he distributed to interested parties subsequent to the 1972 Philosophy of Science Association meeting. Hacking notes that his example uses unbounded utility; apparently I.J. Good pointed out that it rests upon an infinite expected utility. In an earlier publication (Rawling 1994) I mistakenly argue that there is only one "two envelopes problem," the first of the two addressed here.

chance that O contains twice the contents of T. Thus $eav(O)$ is 1.25 times $((0.5 \times 2) + (0.5 \times 0.5))eav(T)$. Thus trading T for O is advantageous, *and* trading O for T is advantageous. Trading is the thing to do, and would remain so even if the identity (O or T) of your envelope were revealed.

2. Two "Two Envelopes" Problems

So runs a variant of the exchange paradox, or "two envelopes problem." I shall argue below that the reasoning in this case is clearly fallacious: it hinges upon algebraic equivocation. I shall then go on, however, to discuss a different variant of the problem. This second variant is really a different problem, and one which raises deeper issues. There are, then, (at least) two different two envelopes problems.

Suppose you know which of the envelopes is T in the circumstance above. Then the choice is effectively that between $10 for sure (envelope O), and a gamble (G) in which a fair coin will be tossed and you will receive $20 if it comes up heads, $5 if tails (envelope T). By standard lights, $eav(G) = [(\$20 + \$5)/2] = \$12.50$, which is greater than $10. Yet the second line of reasoning in the opening paragraph led to the conclusion that $eav(O) = 1.25eav(T) > eav(T)$. It is this second line of reasoning that is fallacious.

The fallacy becomes clear once the reasoning is made explicit. The relevant equation is: $eav(O) = \sum_i p(e_i)v(O \text{ given } e_i)$ where $\{e_i\}$ is a partition of the event space (i.e., exactly one of the e_i will occur), $p(e_i)$ is the probability of e_i, and $v(O \text{ given } e_i)$ is the value of O should e_i transpire. In the case under consideration, the relevant events are:

e_1: the coin came up heads ($v(O \text{ given } e_1) = \$10$; $v(T \text{ given } e_1) = \$20$)

e_2: the coin came up tails ($v(O \text{ given } e_2) = \$10$; $v(T \text{ given } e_2) = \$5$)

and $eav(O) = p(e_1)v(O \text{ given } e_1) + p(e_2)v(O \text{ given } e_2)$, where: $p(e_1) = 0.5$; $p(e_2) = 0.5$; $v(O \text{ given } e_1) = \$10 = 0.5v(T \text{ given } e_1)$; $v(O \text{ given } e_2) = \$10 = 2v(T \text{ given } e_2)$. Thus we have two expressions for $eav(O)$:

$$eav(O) = 0.5(\$10) + 0.5(\$10) = \$10$$

and:

$$eav(O) = 0.5[0.5v(T \text{ given } e_1)] + 0.5[2v(T \text{ given } e_2)] = \$10.$$

The fallacious reasoning incorrectly presumes that:

$$v(T \text{ given } e_1) = v(T \text{ given } e_2) = v(T).$$

It is this false presumption that yields:

$$eav(O) = 0.5[0.5v(T)] + 0.5[2v(T)] = 1.25v(T).$$

It is true that, *given* e_1, $v(O) = 0.5v(T)$, and, given e_2, $v(O) = 2v(T)$. But $v(T)$ differs in the two cases. This is the source of the equivocation.

In the second version of the exchange paradox that I shall consider, it is supposed that you do not know the amount placed in envelope O, although you do know (to simplify matters), that it is $\$2^n$ for some integer $n > 0$. As before, a fair coin was tossed. If it came up heads, $\$2^{n+1}$ were placed in T; if it came up tails, $\$2^{n-1}$ were placed in T. So, where p is your prior probability distribution:

for all $n > 0$:

$$p(T \text{ contains } \$2^{n+1} \mid O \text{ contains } \$2^n) = p(\text{ heads } \mid O \text{ contains } \$2^n) = 0.5$$

$$p(T \text{ contains } \$2^{n-1} \mid O \text{ contains } \$2^n) = p(\text{ tails } \mid O \text{ contains } \$2^n) = 0.5.$$

Now suppose you open what you know to be envelope O (T remains sealed), and find that O contains $\$2^n$ ($n > 0$). You update by conditionalizing on 'O contains $\$2^n$' to yield a posterior distribution, p_N:

$$p_N(\text{heads}) = 0.5 = p_N(\text{tails})$$

where:

$$eav(T \text{ given that } O \text{ contains } \$2^n) = \sum_i p_N(e_i)v(T \text{ given } e_i)$$

the relevant events being:

e_1: the coin came up heads ($v(T \text{ given } e_1) = \$2^{n+1}$)

e_2: the coin came up tails ($v(T \text{ given } e_2) = \$2^{n-1}$).

Thus:

$$eav(T \text{ given that } O \text{ contains } \$2^n) = (0.5 \times \$2^{n+1}) + (0.5 \times \$2^{n-1})$$

$$= 1.25 \times \$2n > \$2n.$$

Alternatively, suppose you open what you know to be envelope T (O remains sealed), and find that T contains $\$2^n$ ($n \geq 0$). Now you must update by conditionalizing on 'T contains $\$2^n$' to yield a posterior distribution, q_N, where:

$$eav(O \text{ given that } T \text{ contains } \$2^n) = \sum_i q_N(e_i)v(O \text{ given } e_i)$$

the relevant events being:

e_1: the coin came up heads ($v(O$ given $e_1) = \$2^{n-1}$ provided that $n >$ 1)

e_2: the coin came up tails ($v(O$ given $e_2) = \$2^{n+1}$).

The difficulty is that

$$p(O \text{ contains } \$2^{n+1} \mid T \text{ contains } \$2^n)$$

(which equals $p(\text{tails} \mid T \text{ contains } \$2^n)$) and

$$p(O \text{ contains } \$2^{n-1} \mid T \text{ contains } \$2^n)$$

(which equals $p(\text{heads} \mid T \text{ contains } \$2^n)$) are as yet undetermined, unless $n = 0$ or 1, in which case:

$$p(\text{tails} \mid T \text{ contains } \$2^n) = 1 \text{ and } p(\text{heads} \mid T \text{ contains } \$2^n) = 0.$$

Suppose $n > 1$. By definition of conditional probability:

$$p(\text{tails} \mid T \text{ contains } \$2^n) = p(\text{tails} \,\&\, T \text{ contains } \$2^n)/p(T \text{ contains } \$2^n).$$

'Tails & T contains $\$2^n$' is equivalent to '$O$ contains $\$2^{n+1}$ & tails.' 'T contains $\$2^n$' is equivalent to '[$O$ contains $\$2^{n+1}$ & tails] or [O contains $\$2^{n-1}$ & heads].' Thus:

$$p(\text{tails} \mid T \text{ contains } \$2^n) =$$
$$\frac{p(O \text{ contains } \$2^{n+1} \,\&\, \text{tails })}{p(O \text{ contains } \$2^{n+1} \,\&\, \text{tails }) + p(O \text{ contains } \$2^{n-1} \,\&\, \text{heads })}.$$

But:

$$p(O \text{ contains } \$2^{n+1} \,\&\, \text{tails }) =$$
$$p(O \text{ contains } \$2^{n+1}) \times p(\text{tails} \mid O \text{ contains } \$2^{n+1}) =$$
$$0.5 \times p(O \text{ contains } \$2^{n+1})$$

and, similarly:

$$p(O \text{ contains } \$2^{n-1} \,\&\, \text{heads }) = 0.5 \times p(O \text{ contains } \$2^{n-1}).$$

Hence:

$$p(\text{tails} \mid T \text{ contains } \$2^n) = \frac{p(O \text{ contains } \$2^{n+1})}{p(O \text{ contains } \$2^{n+1}) + p(O \text{ contains } \$2^{n-1})}.$$

Similarly:

$$p(\text{heads} \mid T \text{ contains } \$2^n) = \frac{p(O \text{ contains } \$2^{n-1})}{p(O \text{ contains } \$2^{n+1}) + p(O \text{ contains } \$2^{n-1})}.$$

The temptation is to suppose that

$$p(\text{tails} \mid T \text{ contains } \$2^n) = p(\text{heads} \mid T \text{ contains } \$2^n) = 0.5,$$

unless $n = 0$ or 1, in which case:

$$p(\text{tails} \mid T \text{ contains } \$2^n) = 1 \text{ and } p(\text{heads} \mid T \text{ contains } \$2^n) = 0.$$

This would yield:

$$q_N(e_1) = q_N(e_2) = 0.5 \text{ if } n > 1$$

$$q_N(e_1) = 0, q_N(e_2) = 1 \text{ if } n = 0 \text{ or } 1$$

which in turn would yield

$$
\begin{aligned}
eav(O \text{ given that } T \text{ contains } \$2^n) &= \sum_i q_N(e_i)v(O \text{ given } e_i) \\
&= (0.5 \times \$2^{n-1}) + (0.5 \times \$2^{n+1}) \\
&= \begin{cases} 1.25 \times \$2^n & \text{if } n > 1 \\ \$2^{n+1} & \text{if } n = 0 \text{ or } 1. \end{cases}
\end{aligned}
$$

In either case $eav(O \text{ given that } T \text{ contains } \$2^n) > \$2^n$. Hence we would have:

for some $n > 0 : O$ contains $\$2^n$;

and for all $n > 0 : eav(T \text{ given that } O \text{ contains } \$2^n) > \$2^n$

and

for some $n \geq 0 : T$ contains $\$2^n$;

and for all $n \geq 0 : eav(O \text{ given that } T \text{ contains } \$2^n) > \$2^n$.

However, the supposition that

$$p(\text{tails} \mid T \text{ contains } \$2^n) = p(\text{heads} \mid T \text{ contains } \$2^n) = 0.5$$

$(n > 1)$ entails that:

for all $n > 1 : 0.5 = \dfrac{p(O \text{ contains } \$2^{n-1})}{p(O \text{ contains } \$2^{n+1}) + p(O \text{ contains } \$2^{n-1})}.$

Hence:

$$\text{for all } n > 1 : p(O \text{ contains } \$2^{n+1}) = p(O \text{ contains } \$2^{n-1}).$$

Suppose this is the case. And suppose, for some integer $m > 0$, $p(O$ contains $\$2^m) = \delta > 0$. Then, for all integers $r > 0$, $p(O$ contains $\$2^{m+2r})$ $= \delta$. For some integer t, $t\delta > 1$. Hence $p([O$ contains $\$2^{m+2}]$ or . . . or $[O$ contains $\$2^{m+2t}]) = t\delta > 1$. For any event A, $1 \geq p(A) \geq 0$. Thus, for all $m > 0$, $p(O$ contains $\$2^m) = 0$. But O contains $\$2^m$ for some $m > 0$, and we have a contradiction if we also suppose countable additivity. A probability measure is countably additive iff:

$$\text{if } A = \bigcup_{i=1}^{\infty} A_i(A_i \cap A_j = \emptyset, i \neq j), \text{ then } p(A) = \sum_{i=1}^{\infty} p(A_i).$$

The event 'O contains $\$2^m$ for some $m > 0$' is the event $\bigcup_{i=1}^{\infty}[O$ contains $\$2^i]$. Thus

$$p[\bigcup_{i=1}^{\infty}[O \text{ contains } \$2^i]] = p[O \text{ contains } \$2^m \text{ for some } m > 0] = 1.$$

But, assuming countable additivity, if for all $i > 0$, $p(O$ contains $\$2^i) = 0$, then

$$p[\bigcup_{i=1}^{\infty}[O \text{ contains } \$2^i]] = \sum_{i=1}^{\infty} p[O \text{ contains } \$2^i] = 0.$$

We must either reject the supposition that

$$p(\text{tails} \mid T \text{ contains } \$2^n) = p(\text{heads} \mid T \text{ contains } \$2^n) = 0.5$$

or reject countable additivity. I address these possibilities in turn. (Of course, if we reject countable additivity, and accept that for all $m > 0$, $p(O$ contains $\$2^m) = 0$, it is neither the case that, for example:

$$\text{for all } n > 1 : 0.5 = \frac{p(O \text{ contains } \$2^{n-1})}{p(O \text{ contains } \$2^{n+1}) + p(O \text{ contains } \$2^{n-1})}$$

nor that

$$\text{for all } n > 0 : p(\text{heads} \mid O \text{ contains } \$2^n) =$$

$$\frac{p(\text{heads } \&O \text{ contains } \$2^n)}{p(O \text{ contains } \$2^n)} = 0.5.$$

See Section 4 for further discussion.)

3. The Countably Additive Framework

Within a countably additive framework, we must reject the supposition that

$$p(\text{tails} \mid T \text{ contains } \$2^n) = p(\text{heads} \mid T \text{ contains } \$2^n) = \frac{1}{2}.$$

The question is: are there alternative (countably additive) probability distributions that yield both:

for some $n > 0 : O$ contains $\$2^n$;

and for all $n > 0 : eav(T$ given that O contains $\$2^n) > \$2^n$

and

for some $n \geq 0 : T$ contains $\$2^n$;

and for all $n \geq 0 : eav(O$ given that T contains $\$2^n) > \$2^n$?

The distribution (distribution A) characterized by the following equations does:

for all $n > 0$: $p(O \text{ contains } \$2^{2n}) = p(O \text{ contains } \$2^{2n-1}) = \dfrac{2^{n-2}}{3^n}$

$p(O \text{ contains } \$x) > 0$ only if $x = 2^n (n > 0)$

for all $n \geq 0$: $p(T \text{ contains } \$2^n) = \dfrac{1}{2}[p(O \text{ contains } \$2^{n-1})+$
$p(O \text{ contains } \$2^{n+1})]$

for all $n > 0$: $p(T \text{ contains } \$2^{n+1} \mid O \text{ contains } \$2^n) = \dfrac{1}{2}$
$= p(T \text{ contains } \$2^{n-1} \mid O \text{ contains } \$2^n)$

for all $n \geq 0$: $p(O \text{ contains } \$2^{n+1} \mid T \text{ contains } \$2^n)$
$+p(O \text{ contains } \$2^{n-1} \mid T \text{ contains } \$2^n) = 1.$

(Compare the distribution in (Broome 1995, pp. 6–7).) Notice that:

$$p(O \text{ contains } \$2) = p(O \text{ contains } \$4) = \frac{1}{6}$$

$$p(O \text{ contains } \$8) = p(O \text{ contains } \$16) = \left(\frac{2}{3}\right)\left(\frac{1}{6}\right)$$

$$p(O \text{ contains } \$32) = p(O \text{ contains } \$64) = \left(\frac{2}{3}\right)\left(\frac{2}{3}\right)\left(\frac{1}{6}\right)$$

etc.

Notice $1 + \left(\dfrac{2}{3}\right) + \left(\dfrac{2}{3}\right)^2 + \left(\dfrac{2}{3}\right)^3 + \cdots = 3$; and $2 \times \left(\dfrac{1}{6}\right) \times 3 = 1$ — i.e.,

$$\sum_1^\infty p(O \text{ contains } \$2^n) = 2\left[\sum_1^\infty \frac{2^{n-2}}{3^n}\right] = 1.$$

$$p(T \text{ contains } \$1) = p(T \text{ contains } \$2) = \left(\frac{1}{2}\right)\left(\frac{1}{6}\right)$$

$$p(T \text{ contains } \$4) = p(T \text{ contains } \$8) = \left(\frac{1}{2}\right)\left[\left(\frac{1}{6}\right) + \left(\frac{2}{3}\right)\left(\frac{1}{6}\right)\right]$$

$$p(T \text{ contains } \$16) = p(T \text{ contains } \$32) =$$
$$\left(\frac{1}{2}\right)\left[\left(\frac{2}{3}\right)\left(\frac{1}{6}\right) + \left(\frac{2}{3}\right)^2\left(\frac{1}{6}\right)\right]$$

etc.

Notice

$$\sum_0^\infty p(T \text{ contains } \$2^n) = 2\left(\frac{1}{2}\right)\left[2\left(\frac{1}{6}\right) + 2\left(\frac{2}{3}\right)\left(\frac{1}{6}\right)\right.$$
$$+ \left. 2\left(\frac{2}{3}\right)^2\left(\frac{1}{6}\right) + \cdots\right] = 2 \times \left(\frac{1}{6}\right) \times 3 = 1.$$

Since, for all $n > 0$:

$$p(T \text{ contains } \$2^{n+1} \mid O \text{ contains } \$2^n) = \frac{1}{2}$$

$$p(T \text{ contains } \$2^{n-1} \mid O \text{ contains } \$2^n) = \frac{1}{2}$$

we have, as above:

$$eav(T \text{ given that } O \text{ contains } \$2^n) = \left(\left(\frac{1}{2}\right) \times \$2^{n+1}\right)$$
$$+ \left(\left(\frac{1}{2}\right) \times \$2^{n-1}\right) = \left(\frac{5}{4}\right) \times \$2^n > \$2^n.$$

In order to calculate $eav(O \text{ given that } T \text{ contains } \$2^n)$, we need to calculate
$$p(O \text{ contains } \$2^{n+1} \mid T \text{ contains } \$2^n).$$

By the definition of conditional probability, this equals:

$$p(T \text{ contains } \$2^n \mid O \text{ contains } \$2^{n+1}) \times \frac{p(O \text{ contains } \$2^{n+1})}{p(T \text{ contains } \$2^n)}.$$

If $n = 0$ or 1, this equals $\left(\dfrac{1}{2}\right) \times \dfrac{(1/6)}{(1/12)} = 1$, and:

$$eav(O \text{ given that } T \text{ contains } \$2^n) = \$2^{n+1} > \$2^n.$$

If $n > 1$, where $p(O \text{ contains } \$2^{n+1}) = \alpha$, it equals:

$$\left(\frac{1}{2}\right) \times \frac{\alpha}{(1/2)((3/2)\alpha + \alpha)} = \frac{\alpha}{(5/2)\alpha} = \left(\frac{2}{5}\right)$$

and

$$eav(O \text{ given that } T \text{ contains } \$2^n) = \left[\left(\frac{2}{5}\right) \times \$2^{n+1}\right]$$

$$+ \left[\left(\frac{3}{5}\right) \times \$2^{n-1}\right] = \left(\frac{11}{10}\right) \times \$2^n > \$2^n.$$

We have a countably additive probability distribution that yields both:

for some $n > 0$: O contains $\$2^n$;

and for all $n > 0$: $eav(T \text{ given that } O \text{ contains } \$2^n) > \$2^n$

and

for some $n \geq 0$: T contains $\$2^n$;

and for all $n \geq 0$: $eav(O \text{ given that } T \text{ contains } \$2^n) > \$2^n$.

Even if you know which envelope you hold (O or T), and how much it contains, you always increase your *eav* by trading for the other envelope (contents unknown, of course). How paradoxical is this?

Nalebuff is certainly correct that "[t]rading envelopes cannot make *both* participants better off" (1989, p. 172) (he envisages two players, each with one of the envelopes). But it does not follow from this that both participants cannot increase their *eav* by trading. Consider the following alternative distribution (distribution B) of the contents of the envelopes:

There are six equally likely pairs, $\langle O, T \rangle$:

$$\langle \$2, \$1 \rangle, \langle \$2, \$4 \rangle, \langle \$4, \$2 \rangle, \langle \$4, \$8 \rangle, \langle \$8, \$4 \rangle, \langle \$8, \$16 \rangle.$$

Note that:

$$eav(T \text{ given that } O \text{ contains } \$8) = \$10 > \$8$$

and

$$eav(O \text{ given that } T \text{ contains } \$4) = \$5 > \$4.$$

Suppose I open my envelope (O) to discover that it contains \$8; and you open your envelope (T) to discover that it contains \$4. We should both trade, since trading increases both our respective *eav*'s. But there is nothing paradoxical about this. Of course, when we trade, I will lose, but that is in the nature of expectation — loss can accompany an increase in *expected* value. (There is a wrinkle here: if I know that you know that you hold T and that you know its contents, and you want to trade, I know that your envelope contains \$4 [you would not want to trade if it contained \$16], hence I would not offer to trade since you will agree iff you win by it. The difficulty is evaded by imposing appropriate ignorance conditions.)

Under distribution B, if you have O, you always increase your *eav* by trading. If you have T, you increase your *eav* by trading iff T contains less than or equal to \$4:

$$eav(O \text{ given that } T \text{ contains less than or equal to } \$4) > \$4;$$

$$eav(O \text{ given that } T \text{ contains more than } \$4) < \$4.$$

If you know that the sums are distributed in accord with B, opening your envelope can provide information vis-à-vis whether or not you increase your *eav* by trading. The odd thing about distribution A, on the other hand, is that, regardless of which envelope you hold, opening your envelope will never affect your trading decision — whichever the envelope and whatever it contains, you will increase your *eav* by trading. Opening your envelope *never* provides information vis-à-vis whether or not you increase your *eav* by trading. Thus you might as well not open it (assuming that you seek to maximize *eav*). But if you do not open it, the rational thing to do, surely, is to compare $eav(O(\text{simpliciter}))$ with $eav(T(\text{simpliciter}))$.

In "standard" cases, such as distribution B, $eav(T) > eav(O)$ (for distribution B, the values are, respectively, \$(35/8) and \$(28/8)). So, if A were a "standard" distribution, the holder of T would receive contradictory advice: on the one hand $eav(T) > eav(O)$, so she should stand pat; on the other, for all relevant n, $eav(O \text{ given that } T \text{ contains } \$2^n) > \$2^n$, so she should trade.

However, A is not "standard": under distribution A, $eav(O)$ and $eav(T)$ are both infinite. This latter is the case for any distribution that suffers the "trading paradox" (for proof, see Nalebuff 1989, pp. 178–179; Castell and Batens 1994, Section 2; Broome 1995, Appendix B; and the appendix to this paper).

To put matters more formally, we would have a contradiction if:

(P) for all $n \geq 0$: $eav(O$ given that T contains $\$2^n) > 2^n$

yet $eav(T) > eav(O)$. For (P) entails that:

for all $n \geq 0 : p(T$ contains $\$2^n)[eav(O$ given that T contains $\$2^n)]$

$$> p(T \text{ contains } \$2^n)\$2^n;$$

and $eav(T) > eav(O)$ entails that:

$$\sum_n p(T \text{ contains } \$2^n)\$2^n (= eav(T)) > \sum_n p(O \text{ contains } \$2^n)\$2^n$$

$$= \sum_n p(T \text{ contains } \$2^n)[eav(O \text{ given that } T \text{ contains } \$2^n)].$$

(the latter are two different expressions for $eav(O)$: $eav(O) = \sum_i p(e_i)v(O$ given e_i); the different expressions are generated by different partitions $\{e_i\}$.) Contradiction is avoided, however, since if (P) holds then $eav(O)$ and $eav(T)$ are both countably infinite.

We are presented with a failure of dominance reasoning in circumstances of infinite expectation. Let $\{e_i\}$ be any partition of the universe into events (note that the partition can be infinite — see Section 4 for further discussion), such that the events are independent of the acts, **f, g** (I assume throughout the sequel that such independence holds; and I will not here explore the issue of causal versus probabilistic independence — see my (1993) for discussion). Then the principle of dominance (PDOM) states that:

(i) if, in each e_i, the agent weakly prefers **f** to **g**, then she should weakly prefer **f** to **g** (weak preference is: strict preference or indifference); and

(ii) if, in each e_i, the agent weakly prefers **f** to **g**, and strictly prefers **f** to **g** in at least one e_i, then she should strictly prefer **f** to **g**.

(In (i), **f** *weakly dominates* **g**; in (ii), **f** *strictly dominates* **g**.) In *eav* terms, PDOM states that:

(i) if $[eav(\mathbf{f}$ given $e_i)] \geq [eav(\mathbf{g}$ given $e_i)]$ for all $e_i \in \{e_i\}$, then $[eav(f)] \geq [eav(g)]$; and

(ii) if $[eav(\mathbf{f}$ given $e_i)] \geq [eav(\mathbf{g}$ given $e_i)]$ for all $e_i \in \{e_i\}$, and $[eav(\mathbf{f}$ given $e_i)] > [eav(\mathbf{g}$ given $e_i)]$ for some $e_i \in \{e_i\}$, then $[eav(\mathbf{f})] > [eav(\mathbf{g})]$.

Obviously, it is possible to re-partition events in such a way that dominance is lost; but if re-partitioning results in a change of preference over acts, or a change in eav(simpliciter) inequalities , then PDOM is violated. And, *a fortiori*, if re-partitioning results in a reversal of strict dominance, then PDOM is violated. For example, either it will rain today, or not, and in either case I would strictly prefer to drive to work rather than take the train: driving strictly dominates taking the train under the partition: {rain, no rain}. However, this is because I believe the chance of heavy traffic to be slight in either case. Driving does not dominate (even weakly) taking the train under the partition:

{(rain & heavy traffic), (no rain & heavy traffic), (rain & traffic not heavy), (no rain & traffic not heavy)}.

But my strict preference for driving over taking the train is maintained (p(rain & heavy traf- fic) and p(no rain & heavy traffic) are both small). (And, *a fortiori*, strict dominance is not reversed). Distribution B furnishes another example: under partitioning with respect to the contents of O, taking T strictly dominates taking O; but under partitioning with respect to the contents of T, taking T does not even weakly dominate taking O. Here are the relevant state/act tables of eav's.

Partitioning with respect to contents of O:

	O holds \$2	O holds \$4	O holds \$8
Take O	\$2	\$4	\$8
Take T	\$(5/2)	\$5	\$10

Taking T strictly dominates taking O: $eav(O) = \$\left[\dfrac{2+4+8}{3}\right] = \$\left(\dfrac{14}{3}\right)$ and $eav(T) = \$\left[\dfrac{(5/2)+5+10}{3}\right] = \$\left(\dfrac{35}{6}\right)$.

Partitioning with respect to contents of T:

	T holds \$1	T holds \$2	T holds \$4	T holds \$8	T holds \$16
Take O	\$2	\$4	\$5	\$4	\$8
Take T	\$1	\$2	\$4	\$8	\$16

Neither option dominates: $eav(O) = \$\left[\left(\dfrac{2}{6}\right) + \left(\dfrac{4}{6}\right) + \left(\dfrac{5}{3}\right) + \left(\dfrac{4}{6}\right) + \left(\dfrac{8}{6}\right)\right] = \$\left(\dfrac{14}{3}\right)$ and $eav(T) = \$\left[\left(\dfrac{1}{6}\right) + \left(\dfrac{2}{6}\right) + \left(\dfrac{4}{3}\right) + \left(\dfrac{8}{6}\right) + \left(\dfrac{16}{6}\right)\right]$

$= \$ \left(\dfrac{35}{6} \right)$. Repartitioning does not alter *eav*'s (simpliciter). Thus, given that you do not know the contents of either envelope, repartitioning does not alter your preference for T over O.

In the "trading paradox," on the other hand, re-partitioning results in strict dominance reversal. We have:

for all $n \geq 0$: $[eav(O \text{ given that } T \text{ contains } \$2^n)] > \$2^n$

for all $n > 0$: $[eav(T \text{ given that } O \text{ contains } \$2^n)] > \$2^n$

i.e., under the partition of events:

$$\{T \text{ contains } \$2^i : i \text{ an integer}, i \geq 0\},$$

taking O strictly dominates taking T. But under the partition:

$$\{O \text{ contains } \$2^i : i \text{ an integer}, i > 0\},$$

taking T strictly dominates taking O. Thus, by PDOM, $eav(O) > eav(T)$ *and* $eav(T) > eav(O)$. In actual fact, $eav(O) = eav(T)$, both being (countably) infinite.

The two envelopes puzzle presents us, then, with a violation of dominance reasoning in circumstances of infinite expectation. That dominance reasoning is violated in such circumstances is no surprise: there are plenty of less subtle cases. Compare the following pair of gambles. In both, a fair coin is tossed until tails occurs, and you receive a prize dependent upon the number of tosses, n. In the first gamble (G_1), you receive a prize of $\$2^n$; in the second ($G_2$) you receive a prize of $\$2^{n+1}$. G_2 strictly dominates G_1. Yet their *eav*'s $\sum\limits_{1}^{\infty} (2^n)(2^{-n})$ and $\sum\limits_{1}^{\infty} 2(2^n)(2^{-n})$, respectively, are both countably infinite. Thus, if infinite expectations are entertained, we have two acts of equal *eav*, one of which strictly dominates the other. (For other examples of the breakdown of decision theory in the face of infinite expectation, see (Jeffrey 1983, pp. 153–154.))

The trading "paradox" is that, whichever envelope you hold, and whatever it contains, you increase your *eav* by trading. In "standard" cases, if you hold O and reason that you increase your *eav* by trading no matter what it contains, this is because:

either you reason that, under the partition of events: $\{O \text{ contains } \$2^i : i \text{ an integer}, i \geq 0\}$, T strictly dominates O with respect to *eav*; or you reason that $eav(O) > eav(T)$

and it does not matter which line of reasoning you pursue, because PDOM guarantees that:

if: for all $n \geq 0$: $[eav(T \text{ given that } O \text{ contains } \$2^n)] > \$2^n$

then: $eav(T) > eav(O)$.

But PDOM breaks down in the face of infinite expectations. We have:

for all $n \geq 0$: $[eav(T \text{ given that } O \text{ contains } \$2^n)] > \$2^n$

and $eav(O) = eav(T)$ (both being countably infinite).

Admittedly (as John Broome remarked to me in conversation), the case of G_1 versus G_2 above is less puzzling than the trading paradox. Even though it illustrates a breakdown of PDOM in the face of infinite expectations, Broome would certainly prefer G_2, regardless of whether he currently holds either of the gambles. In the case of the envelopes, if he holds O, he prefers T; if he holds T, he prefers O.

Castell and Batens (1994, p. 49) claim that, "the paradox derives from the fact that if a random variable has infinite expectation, then its true value is bound to be less than its expectation value." I demur. If Castell and Batens were correct, then infinitude would be sufficient for the "paradox," and it is not. The trading argument is not driven by the prospect of exchanging the finite for the infinite: both envelopes contain finite sums. The case of G_1 versus G_2 is a case in point. Suppose O contains the winnings of G_1, T the winnings of G_2, then $eav(O)$ and $eav(T)$ are both infinite, yet the trading paradox does not arise:

for all $n > 0$: $eav(O \text{ given that } T \text{ contains } \$2^{n+1}) = \$2^n < \$2^{n+1}$

for all $n > 0$: $eav(T \text{ given that } O \text{ contains } \$2^n) = \$2^{n+1} > 2^n$.

Taking T strictly dominates taking O under the partition:

$$\{T \text{ contains } \$2^{i+1} : i \text{ an integer}, i > 0\}$$

and under the partition:

$$\{O \text{ contains } \$2^i : i \text{ an integer}, i > 0\}.$$

Several authors note (see, e.g., Nalebuff 1989, p. 178; Broome 1995, p. 8) that the trading paradox is akin to the St. Petersburg paradox in that both are centered upon infinite expectation. Some authors see the latter paradox as the fact that, despite their infinite expected utilities, "no one would be willing to pay very much" (Broome 1995, p. 9) for the opportunity of being exposed to G_1 or G_2, their infinite eav's notwithstanding. Other authors see the St. Petersburg paradox as an argument leading to a problem for any decision theory which admits of unbounded finite utilities:

The St. Petersburg paradox is the argument that, given an infinite sequence of consequences whose desirabilities are finite but have no finite upper bound, we can use the St. Petersburg construction to prove the existence of a gamble of infinite desirability (Jeffrey 1983, pp. 152–153).

That is, given prizes of unbounded but finite utility, prizes can be found of value at least 2^n utiles for each n, and hence a gamble that dominates G_1 (make the prizes utiles rather than dollars) can be constructed. (This argument is due to Bernoulli, 1738, who first discussed the St. Petersburg game. See Jeffrey 1983, p. 22.) One solution is to bound utility. But Jeffrey offers an alternative. In his *Logic of Decision*, the preference ranking comprises propositions in which the agent has positive degree of belief, and since no rational agent has positive degree of belief that any mortal, or any secular institution (I leave aside religious and supernatural possibilities here), can offer her prizes of arbitrarily large, albeit finite, desirability, St. Petersburg gambles do not appear in the preference ranking: "Put briefly and crudely, our rebuttal of the St. Petersburg paradox consists in the remark that anyone who offers to let the agent play the St. Petersburg game is a liar, for he is pretending to have an indefinitely large bank" (Jeffrey 1983, p. 154).

In the countably additive case, what drives the "paradox" is the inequality (see appendix):

$$\text{for all } n \geq 0 : p(O \text{ contains } \$2^{n+1}) > \left(\frac{1}{2}\right) p(O \text{ contains } \$2^{n-1}).$$

Anyone who offers you an envelope containing a sum determined by a chance mechanism satisfying this inequality is also "pretending to have an indefinitely large bank." Thus the pair of envelopes presents no problem for *The Logic of Decision*: the rational agent has zero degree of belief in the proposition that the sums in the envelopes are distributed in such a way as to give rise to the "trading paradox," and so the relevant propositions describing the envelopes do not appear in her preference ranking. On this account, the response to someone who is puzzled by the fact that if she holds O, she prefers T; if she holds T, she prefers O — so that she could see herself trading back and forth (assuming she does not open either of the envelopes) *ad infinitum* (and perhaps paying a small sum for each trade, and thereby being "pumped" out of her fortune) — is to point out that she has been hoodwinked into believing a distribution no individual can offer her.

Jeffrey's argument is actually stronger than we require: it does not hinge upon infinite expectation. It works against, for example, the following distribution:

$$p(O \text{ contains } \$1) = \frac{2}{3}$$

for all integers $n > 0$: $p(O \text{ contains } \$2^n) = \left(\dfrac{1}{4}\right)^n$

Yet:

$$\sum_0^\infty [2^i p(O \text{ contains } \$2^i)] = \frac{5}{3}.$$

The finite expectation notwithstanding, it is still the case that, for any n, there is a chance that anyone who offers you an envelope containing a sum determined by a chance mechanism satisfying this distribution will have to come up with more than $\$2^n$. Even if someone uses the appropriate mechanism, and, say, the most likely outcome occurs, so that she only has to place $1 in the envelope, she must still have been able, if her offer were genuine, to come up with arbitrarily large sums. And this she could not do.

In theories in which utility is bounded, appeal is made to the diminishing marginal utility of money to ensure conformity to the bound: expected utility is not linearly related to *eav* beyond a certain point. And this diminishing marginal utility of money vitiates the reasoning underlying the "trading paradox." Suppose the agent's envelope contains M, then (where EU is expected utility, U utility):

$EU(\text{Trade}) = U(2M)p(2M \mid \text{Trade}) + U(M/2)p((M/2) \mid \text{Trade}).$

Suppose:

$$[(2M)p(2M \mid \text{Trade}) + (M/2)p((M/2) \mid \text{Trade})] > M$$

then:

$$[2U(M)p(2M \mid \text{Trade}) + (1/2)U(M)p((M/2) \mid \text{Trade})] > U(M).$$

Suppose further that:

$$U(nM) = n[U(M)]$$

then:

$$EU(\text{Trade}) > U(M).$$

However, the diminishing marginal utility of money entails that for large M: $U(2M) < 2U(M)$, hence it is possible that $U(M) \geq EU(\text{Trade})$.

I have been assuming that individuals are proffering the envelopes, and that they cannot possess arbitrarily large sums of money. But perhaps nations can. However, as Jeffrey (1983, p. 155) notes, diminishing marginal utility of money would certainly apply if a nation were to print enormous denomination bills to pay you: consider the resulting inflation . . .

Due to their common concern with infinite expected utilities, both the St. Petersburg paradox and the two envelopes "trading paradox" (under countable additivity) can be undercut by evading infinitude (see also, e.g., Broome 1995, p. 9). However, whereas the St. Petersburg paradox is an argument that takes us from the finite to the infinite, the two envelopes problem is one illustration, amongst many, of the difficulties attendant upon our arrival.

4. The Rejection of Countable Additivity

The alternative to countable additivity is to insist only that probability be finitely additive. We insist only that:

$$\text{if } A = \sum_{i=1}^{n} A_i, \text{ some } n(A_i \cap A_j = \emptyset, i \neq j), \text{ then } p(A) = \sum_{i=1}^{n} p(A_i)$$

and drop the additional requirement that:

$$\text{if } A = \bigcup_{i=1}^{\infty} A_i(A_i \cap A_j = \emptyset, i \neq j), \text{ then } p(A) = \sum_{i=1}^{\infty} p(A_i).$$

De Finetti (1972, p. 105) is "convinced that [countable] additivity cannot really be defended" (see, for example, chapter five of his (1972) for some of his reasons). And Savage's (1972) axiomatization of decision theory "does not imply countable additivity, but only finite additivity" (p. 40). I agree with Savage that "many of us have a strong intuitive tendency to regard as natural probability problems about the necessarily only finitely additive uniform probability densities on the integers, on the line, and on the plane" (1972, p. 43; see pp. 40–43 for further discussion of finite versus countable additivity). In the case of, for instance, the uniform probability density on the integers:

$$\text{for all integers } n : p(n \text{ is chosen}) = 0$$

(if for all integers n: $p(n \text{ is chosen}) = \delta > 0$, then there is m such that $m\delta > 1$, and $\sum_{n=1}^{m} p(n \text{ is chosen}) > 1$) yet $p(\text{some } n \text{ is chosen}) = 1$.

The exchange paradox was perhaps first introduced to the philosophical community at the biennial meeting of the Philosophy of Science Association in 1972 by Ian Hacking (his contribution to that meeting was unfortunately not published in the Proceedings). He developed an example of Dennis Lindley's (1972, p. 50).

Lindley's example supposes only bounded utility, and requires that probability be only finitely additive. Let a_m be an increasing function everywhere greater than zero, and bounded above at one (m is an integer, positive, negative or zero). There are two cards, face down. You know that one card is marked with an integer, m, chosen at random (i.e., we have a uniform prior on the integers; hence the requirement of mere finite additivity); the other with $m + 1$ (you know not which is which). You have agreed to the following gamble. You will pick a card. If it is marked with the larger of the two integers, $m + 1$, you will receive a_{m+1} utiles; if, on the other hand, it is marked with the smaller of the two, m, you will lose a_m utiles. But now you are in a quandary. Designate the integer on the left-hand card 'q'; that on the right-hand card 'r.' We have the following pair of equations:

for all n: $p(r = n + 1 \mid q = n) = p(r = n - 1 \mid q = n) = \dfrac{1}{2}$

for all n: $p(q = n + 1 \mid r = n) = p(q = n - 1 \mid r = n) = \dfrac{1}{2}$.

Under the partition, $\{q = n : n \text{ an integer}\}$:

$$\text{for all } n : EU(\text{right} \mid q = n) \;=\; \frac{1}{2}(-a_{n-1} + a_{n+1})$$
$$> \;\; EU(\text{left} \mid q = n) = \frac{1}{2}(a_n - a_n) = 0.$$

(If you have the left-hand card, you want to trade for the right-hand.) But under the partition, $\{r = n : n \text{ an integer}\}$:

$$\text{for all } n : EU(\text{left} \mid r = n) \;=\; \frac{1}{2}(-a_{n-1} + a_{n+1})$$
$$> \;\; EU(\text{right} \;\; r = n) = \frac{1}{2}(a_n - a_n) = 0.$$

(If you have the right-hand card, you want to trade for the left-hand.) Thus, absent countable additivity and with bounded utility, dominance with respect to countably infinite partitions is violated, and we have a "trading paradox."

Hacking modified Lindley's example to conform to countable additivity, and arrived at the problem that takes center stage in Section 3 (he points out the violation of dominance). Hence his case requires that we countenance infinite utility.

Lindley's example involves infinitely many consequences, and requires conditioning on events of probability zero (for all n: $p(r = n) = p(q =$

$n) = 0$). Compare the case of the envelopes, if we countenance both:

$$\text{for all } n > 0 : p(T \text{ contains } \$2^{n+1} \mid O \text{ contains } \$2^n)$$

$$= p(T \text{ contains } \$2^{n-1} \mid O \text{ contains } \$2^n) = 0.5$$

and

$$\text{for all } n > 1 : p(O \text{ contains } \$2^{n+1} \mid T \text{ contains } \$2^n)$$

$$= p(O \text{ contains } \$2^{n-1} \mid T \text{ contains } \$2^n) = 0.5.$$

We derived:

$$\text{for all } n > 1 : p(O \text{ contains } \$2^{n+1}) = p(O \text{ contains } \$2^{n-1}).$$

In order to do this, as it turns out, we accepted such absurdities as: $0/0 = 0.5$. We allowed ourselves to condition on events of probability zero, under the "standard" definition of conditional probability:

$$p(E|H) = [p(E\&H)/p(H)].$$

However, there are alternative accounts on which conditioning on events of probability zero is acceptable (see, e.g., de Finetti 1972, Chapter 5). Suppose we stipulate as one of the initial conditions of the two envelopes problem that the amount in O is chosen at random from the set: $\{\$2n : n \text{ an integer greater than } 0\}$. This entails that:

$$\text{for all } n > 0 : p(O \text{ contains } \$2^n) = 0.$$

We then condition on events of probability zero to generate the "trading paradox" — we have both:

$$\text{for all } n > 0 : p(T \text{ contains } \$2^{n+1} \mid O \text{ contains } \$2^n)$$

$$= p(T \text{ contains } \$2^{n-1} \mid O \text{ contains } \$2^n) = 0.5$$

and

$$\text{for all } n > 1 : p(O \text{ contains } \$2^{n+1} \mid T \text{ contains } \$2^n)$$

$$= p(O \text{ contains } \$2^{n-1} \mid T \text{ contains } \$2^n) = 0.5.$$

Within the finitely additive framework, then, we have seen two cases of trading paradoxes: Lindley's example, and the envelopes. The envelopes problem still suffers the objection that no one can offer you an amount of money chosen at random from $\{\$2^n : n \text{ an integer greater than } 0\}$. Lindley's example, however, does not involve unbounded utilities, although it does involve infinitely many consequences. Both examples involve conditioning on events of probability zero and violations of infinite-partition dominance.

There are also examples of violation of infinite-partition dominance in the absence of countable additivity that (i) require neither unbounded utility nor infinitely many consequences; and (ii) do not require conditioning upon events of probability zero. Consider the following example of de Finetti's (1972, pp. 99–100).

Let A_1, B_1, A_2, B_2, A_3, . . . be mutually disjoint events.

$A = \cup A_i$; $B = \cup B_i$; $A \cup B = $ the sure event.

For all i, $p(A_i) = p(B_i) = p_i > 0$; $\sum p_i = 1/3$ (countable additivity is rejected);

$p(A) = 1/3$; $p(B) = 2/3$.

Note that, for all i, $p(A_i \mid A_i \cup B_i) = (p_i/2p_i) = 1/2$.

Consider the following pair of gambles (Seidenfeld and Schervish 1983, pp. 406–407):

G_1: Win \$1.50 if B, lose \$1.50 if A.

G_2: Win \$2 if A, lose \$1 if B.

Since $p(A) = 1/3$; $p(B) = 2/3$:

$$eav(G_1) = \sum p(e_i)v(G_1 \text{ given } e_i) = \frac{2}{3} \times \$1.50 - \frac{1}{3} \times \$1.50$$

$$= \$0.50 > \$0 = eav(G_2).$$

Thus G_1 is strictly preferred to G_2. But, for all i, $p(A_i \mid A_i \cup B_i) = 1/2$; hence, for all i:

$$eav(G_2 \mid A_i \cup B_i) = \$0.50 > \$0 = eav(G_1|A_i \cup B_i).$$

Thus, by PDOM, G_2 is strictly preferred to G_1 . . . Note further that:

$$eav(G_1) = \sum p(e_i)v(G_1 \text{ given } e_i) = \$1.50 \times \sum p(B_i) - \$1.50 \times p(A_i) = \$0$$

$$eav(G_2) = \$2 \times \sum p(A_i) - \$1 \times \sum p(B_i) = \$(1/3).$$

With infinite partitions, matters can go awry in the finitely additive framework with respect to dominance reasoning and expectation.

Savage's (1972) axiomatization entails bounded utility (see, e.g., Fishburn 1981, p. 161); and it entails that: event A is of probability zero iff for all acts $\mathbf{f}$, $\mathbf{g}$, $EU(\mathbf{f}$ given $A) = EU(\mathbf{g}$ given $A)$ (Savage 1972, p. 24). This latter feature entails that Lindley's example does not square with Savage's theory. De Finetti's example, however, shows that Savage cannot derive even the following dominance principle (which is weaker than the usual principle: note the strict '>' in the antecedent) with respect to *infinite* partitions $\{e_i\}$:

$$\text{if } [EU(\mathbf{f} \text{ given } e_i)] > [EU(\mathbf{g} \text{ given } e_i)] \text{ for all}$$

$$\text{events } e_i \in \{e_i\}, \text{ then } [EU(\mathbf{f})] \geq [EU(\mathbf{g})].$$

Seidenfeld and Schervish (1983) discuss this point in detail; and they offer illuminating more general discussion of both de Finetti's and Savage's approaches.

Whereas in the countably additive case, dominance reasoning breaks down in the face of infinite expectation, in the merely finitely additive framework, infinite partitions can vitiate not only dominance reasoning, but also the very notion of expectation (expected utility is no longer guaranteed even to be a function). The two envelopes problem is a relevant example: O dominates T with respect to one infinite partition; T dominates O with respect to another; yet the *eav* of each envelope appears to be zero (e.g., $eav(O) = \sum_n p(O \text{ contains } \$2^n)\$2^n = \sum_n 0 \times \$2^n = 0$). Perhaps no one can offer you the envelopes because of financial limitations, but this objection does not apply to the development of de Finetti's example above. If probability is merely finitely additive, it seems we must abandon decision theoretic reasoning with respect to infinite partitions.

Appendix

Theorem. The conjunction of:

(1) $p(T \text{ contains } 2^{n+1} \mid O \text{ contains } 2^n) = (1/2) = p(T \text{ contains } 2^{n-1} \mid O \text{ contains } 2^n)$, for all $n > 0$;

and

(2) $p(O \text{ contains } 2^{n+1} \mid T \text{ contains } 2^n) + p(O \text{ contains } 2^{n-1} \mid T \text{ contains } 2^n) = 1$, for all $n \geq 0$

entails that $eav(O$ given that T contains $2^n) > 2^n$, all $n \geq 0$, iff:

(∗) $p(O$ contains $2^{n+1}) > (1/2)p(O$ contains $2^{n-1})$, all $n \geq 0$.

Proof. By (2):

$$\text{for all } n \geq 0 : eav(O \text{ given that } T \text{ contains } 2^n)$$
$$= 2^{n+1}p(O \text{ contains } 2^{n+1} \mid T \text{ contains } 2^n)$$
$$+2^{n-1}p(O \text{ contains } 2^{n-1} \mid T \text{ contains } 2^n).$$

So:

$eav(O$ given that T contains $2^n) > 2^n$, all $n \geq 0$, iff

$$2p(O \text{ contains } 2^{n+1} \mid T \text{ contains } 2^n)$$
$$+\frac{1}{2}p(O \text{ contains } 2^{n-1} \mid T \text{ contains } 2^n) > 1, \text{ all } n \geq 0.$$

Thus, by (2):

$[eav(O$ given that T contains $2^n) > 2^n$, all $n \geq 0$, iff

$p(O$ contains $2^{n+1} \mid T$ contains $2^n) > \frac{1}{3}$, all $n \geq 0$.

Also, by (2):

$$\text{for all } n \geq 0 : p(T \text{ contains } 2^n) = p(T \text{ contains } 2^n \& O \text{ contains } 2^{n-1})$$
$$+ p(T \text{ contains } 2^n \& O \text{ contains } 2^{n+1}).$$

So, by (1):

$$\text{for all } n \geq 0 : p(T \text{ contains } 2^n) = (1/2)[p(O \text{ contains } 2^{n-1})$$
$$+p(O \text{ contains } 2^{n+1})].$$

Hence, by (1), (2):

$$\text{for all } n \geq 0 : p(O \text{ contains } 2^{n+1} \mid T \text{ contains } 2^n)$$
$$= \frac{1}{2}\frac{p(O \text{ contains } 2^{n+1})}{p(T \text{ contains } 2^n)} = \frac{p(O \text{ contains } 2^{n+1})}{p(O \text{ contains } 2^{n-1}) + p(O \text{ contains } 2^{n+1})}.$$

Therefore $eav(O$ given that T contains $2^n) > 2^n$, all $n \geq 0$, iff:

(∗) $p(O$ contains $2^{n+1}) > (1/2)p(O$ contains $2^{n-1})$, all $n \geq 0$.

Theorem.

(P) for all $n \geq 0$: $eav(O$ given that T contains $2^n) > 2^n$

entails that $eav(O)$ and $eav(T)$ are both infinite.

Proof. Recall that, assuming (1) and (2), (P) holds iff:

$(*)$ $p(O$ contains $2^{n+1}) > (1/2)p(O$ contains $2^{n-1})$, for all $n \geq 0$.

Suppose $(*)$. Note that $eav(O) = \sum_{1}^{\infty}[2^i p(O$ contains $2^i)]$. Let $r_n = \sum_{1}^{n}[2^i p(O$ contains $2^i)]$. Thus $eav(O) = \lim_{n\to\infty} r_n$. Now suppose, for *reductio*, that $\lim_{n\to\infty} r_n = m$, some finite m. For i an integer > 0: $r_i = m - \delta_i$; where $\delta_i > 0$, $\lim_{i\to\infty} \delta_i = 0$. Now

$$r_{i+2} = m - \delta_{i+2} = r_i + 2^{i+1}p(O \text{ contains } 2^{i+1}) + 2^{i+2}p(O \text{ contains } 2^{i+2}).$$

By $(*)$,

$$2^{i+1}p(O \text{ contains } 2^{i+1}) > 2(2^{i-1})p(O \text{ contains } 2^{i-1}), \text{ all } i \geq 0.$$

Thus $r_{i+2} > 2r_i$, and hence $m - \delta_{i+2} > 2m - 2\delta_i$. $\delta_{i+2} > 0$, so $\delta_i > (m/2)$. Therefore $\lim_{i\to\infty} \delta_i \neq 0$ and $\lim_{n\to\infty} r_n \neq m$. Note that $eav(O)$ is infinite iff $eav(T)$ is infinite. (Let $w_n = \sum_{0}^{n}[2^i p(T$ contains $2^i)]$. Recall that $p(O$ contains $x) > 0$ only if $x = 2^n$ $(n > 0)$; and that, for all $n \geq 0$, $p(T$ contains $2^n) = \frac{1}{2}[p(O$ contains $2^{n-1}) + p(O$ contains $2^{n+1})]$. So:

$$eav(T) = \lim_{n\to\infty} w_n = \sum_{0}^{\infty}\left[2^i \left(\frac{1}{2}\right)[p(O \text{ contains } 2^{i-1})\right.$$

$$\left. + p(O \text{ contains } 2^{i+1})]\right] = \frac{5}{4}\sum_{1}^{\infty}[2^i p(O \text{ contains } 2^i)].$$

REFERENCES

Broome, J. (1995). The Two-Envelope Paradox. *Analysis* **55**, 6–11.

Castell, P. and D. Batens (1994). The Two Envelope Paradox: The Infinite Case. *Analysis* **54**, 46–49.

de Finetti, B. (1972). *Probability, Induction and Statistics.* New York: Wiley.

Fishburn, P. (1981). Subjective Expected Utility: A Review of Normative Theories. *Theory and Decision* **13**, 139–199.

Lindley, D. (1972). *Bayesian Statistics: A Review.* Philadelphia: Society for Industrial and Applied Mathematics.

Nalebuff, B. (1989). The Other Person's Envelope is Always Greener. *Journal of Economic Perspectives* **3**, 171–181.

Jeffrey, R. (1983). *The Logic of Decision,* 2nd ed. Chicago: University of Chicago Press.

Rawling, P. (1993). Choice and Conditional Expected Utility. *Synthese* **94**, 303–328.

Rawling, P. (1994). A Note on the Two Envelopes Problem. *Theory and Decision* **36**, 97–102.

Savage, L. (1972). *The Foundations of Statistics,* 2nd ed. New York: Dover.

Seidenfeld, T. and M. Schervish (1983). A Conflict Between Finite Additivity and Avoiding Dutch Book. *Philosophy of Science* **50**(3), 398–412.

Piers Rawling
Department of Philosophy
University of Missouri — St. Louis
St. Louis, Missouri 63121-4499
e-mail: sjprawl@umslvma.umsl.edu

Commentary by Robert Gardner

Dr. Rawling states two versions of the two envelopes problem. In both versions, an amount of money is placed in an envelope O, a coin is tossed and twice the amount of money in O is placed in envelope T if the coin comes up heads or half the amount of money in O is placed in envelope O is placed in envelope T if the coin comes up tails. As he observes, if the amount of money in O is \$10 (or any known quantity x), then the expected amount in T is \$1.25 × 10 (or \$1.25x in general). Therefore, in such a situation no "exchange paradox" arises. However, if the amount of money in O is not known (say it is some value 2^n where $n > 0$ is an integer), then he argues that this leads to the exchange paradox in which if one holds envelope O, then s/he deduces that the expected amount in envelope T is greater, and conversely if one holds envelope T, then s/he deduces that the expected amount in evelope O is greater. Since this is clearly contradictory, Dr. Rawling is lead to the conclusion that he must either reject

1. P(tails | T contains \$$2^n$) = P(heads | T contains \$$2^n$) = 0.5, or

2. countable additivity.

I propose that a third option exists to resolve the paradox which is perhaps even more elementary.

    If we interpret the two envelopes problem to consist of the following events (in order):

1. put an amount of money x in envelope O,

2. flip a coin,

3. put twice the amount x in envelope T if the coin comes up heads, and put half the amount x in envelope T if the coin comes up tails,

then the exchange paradox is easily explained. The real problem lies in determining *how* the amount x which is to be placed in envelope O is to be determined. If we are given a probability distribution that describes how the quantity is chosen, then the paradox immediately dissappears. Suppose this probability distribution has mean \$$\mu$. Then the expected amount in envelope O is \$$\mu$ and the expected amount in T is \$$\frac{5}{4}\mu$. No paradox exists and one should choose envelope T. Notice this can be accomplished over a countable sample space for which the probability of no event is zero and yet we still have countable additivity and finite expected values, as illustrated in the following example.

Example 1. Suppose a positive integer n is chosen according to the probability distribution $p(n) = \dfrac{1}{2^n}$ and \$$\left(\dfrac{3}{2}\right)^n$ is placed in envelope O. A coin is tossed and an amount placed in envelope T as described above. Then the expected amount in O is $\displaystyle\sum_{n=1}^{\infty} \left(\frac{3}{2}\right)^n \left(\frac{1}{2^n}\right) = 3$ and the expected amount in T is $1.25 \times 3 = \dfrac{15}{4}$. It is fallacious to argue that the expected value of O is $\dfrac{5}{4}$ times the expected value of O, since the expected value of O is already determined by the probability distribution.

Of course, one can argue that the game given in Example 1 cannot actually be played since it requires the availability of an unbounded amount of money. In the following example, this problem does not arise.

Example 2. Suppose a positive number between 0 and 1 is chosen according to a uniform distribution (i.e. the probability function if $f(x) = 1$) is placed in envelope O. Then the probability that a value from set $A \subset (0,1)$ is chosen is $\int_A 1$ (for generality, we take the integration to be Lebesgue integration). The expected amount in O is $\int_{(0,1)} x = \frac{1}{2}$. Following the coin toss as described above, the expected amount in envelope T is $\frac{5}{8}$. Again, there is no paradox and one should choose envelope T. Notice that in this example, we have countable additivity (since Lebesgue integration is countably additive), although the sample space is uncountable.

Notice that one can argue that $eav(T) = \frac{5}{4} eav(O)$. However, as above, it is not valid to calculate $eav(O)$ in terms of $eav(T)$ as $eav(O) = \frac{e}{4} eav(T)$, since $eav(O)$ is given by the probability distribution. In addition, we still have $p(\text{tails} \mid T \text{ contains } x) = p(\text{heads} \mid T \text{ contains } x) = 0.5$.

Robert B. Gardner
Department of Mathematics
Department of Physics and Astronomy
East Tennessee State University
Box 70663
Johnson City, TN 37614
e-mail: gardnerr@etsu.edu

Rawling's Reply to Gardner

I thank Professor Gardner for his commentary. I shall here respond to him on two issues, largely for purposes of clarification.

Professor Gardner suggests that

(A) $p(\text{tails} \mid T \text{ contains } \$2n) = p(\text{heads} \mid T \text{ contains } \$2n) = 0.5$

is consistent with countable additivity. I demur, given the proviso (which is an initial condition of the problem):

(B) $p(\text{tails} \mid O \text{ contains } \$2n) = p(\text{heads} \mid O \text{ contains } \$2n) = 0.5$.

(A) and (B) (for $n > 1$ and $n > 0$, respectively) together entail a uniform distribution of the contents of O over infinitely many integers (and this clearly contravenes countable additivity, given that O contains \$2n for some n).

Here is an abbreviated version of the proof I give of the claim:

"T contains \$2n" is equivalent to
"[O contains \$$(2n + 1)$ & tails] or [O contains \$$(2n - 1)$ & heads]"

Thus:

$$p(\text{tails} \mid T \text{ contains } \$2n)$$

$$= \frac{p(O \text{ contains } \$(2n + 1) \, \& \text{ tails})}{p(O \text{ contains } \$(2n + 1) \, \& \text{ tails}) + p(O \text{ contains } \$(2n - 1) \, \& \text{ heads})}.$$

But:

$$p(O \text{ contains } \$(2n + 1) \, \& \text{ tails}) = p(O \text{ contains } \$(2n + 1)) \times$$

$$p(\text{tails} | O \text{ contains } \$(2n + 1)) = 0.5 p(O \text{ contains } \$(2n + 1)) (\text{by (B)})$$

and, similarly:

$$p(O \text{ contains } \$(2n - 1) \, \& \text{ heads}) = 0.5 p(O \text{ contains } \$(2n - 1)).$$

Hence:

$$p(\text{tails} \mid T \text{ contains } \$2n)$$

$$= \frac{p(O \text{ contains } \$(2n + 1))}{p(O \text{ contains } \$(2n + 1)) + p(O \text{ contains } \$(2n - 1))}.$$

Thus, by (A):

$$\text{for all } n > 1 : 0.5 = \frac{p(O \text{ contains } \$(2n + 1))}{p(O \text{ contains } \$(2n + 1)) + p(O \text{ contains } \$(2n - 1))}.$$

Hence:

$$\text{for all } n > 1 : p(O \text{ contains } \$(2n + 1)) = p(O \text{ contains } \$(2n - 1)).$$

Professor Gardner claims that there is no exchange paradox given countable additivity, provided that the mean of the distributed amount in O is finite. I agree. Here is a summary of part of my paper.

Whilst a probability distribution satisfying (A) and (B) above is ruled out in the countably additive framework, there are probability distributions that are consistent with countable additivity and give rise to an exchange "paradox." But it transpires that if such a probability distribution holds, then the mean of the distributed amount in O is infinite.

Does this entailment resolve the problem? It is certainly the case that, under such a distribution, $eav(O) = eav(T)$, both being infinite. But the "paradox" still perhaps obtains. Given that you hold an envelope, its contents are finite — say m. And the eav of the other envelope, conditional upon yours containing m, is greater than m.

Is this really so paradoxical, however? One suggestion is to claim that matters only appear strange because they are based upon the impossible supposition that someone could genuinely offer you a sum selected according to a distribution with an infinite mean.

Poznań Studies in the Philosophy of the Sciences and the Humanities
2000, *vol.* 71, *pp.* 77–94

Susan Vineberg

THE LOGICAL STATUS OF CONDITIONALIZATION AND ITS ROLE IN CONFIRMATION [1]

ABSTRACT. The rule of conditionalization plays a prominent role in the Bayesian theories of confirmation and decision. Although various justifications for the principle have been proposed, the most widely discussed has been the so-called Dutch Strategy argument. In the first part of the paper I argue, contrary to several recent authors, that the Dutch Strategy argument does reveal something important about the status of the conditionalization rule, but that it does not establish the strong version of the principle of conditionalization that it has sometimes been promoted as justifying. In the second part of the paper the role of the conditionalization rule in confirmation will be discussed.

1. The Principle of Conditionalization

The principle of conditionalization is generally stated as requiring that an agent's new probability[2] for A after learning E, and nothing more, should be equal to her old probability of A given E. If p' is the agent's new probability function after learning E and nothing stronger, and p her old probability function, then the principle requires that for each proposition A in the domain of p, $p'(A) = p(A|E)$, where $p(A|E)$ is the prior probability of A given E. By definition, $p(A|E) = p(A\&E)/p(E)$. So stated, the principle of conditionalization can be thought of as demanding that upon updating the probability of E to one, that the probability of each proposition always be changed by the rule of conditionalization, which involves setting the new probability for each proposition equal to its old probability conditional on the proposition learned.

If the principle of conditionalization is adopted, then future opinions are determined by a strict formula involving current opinion together with the new information obtained. Opinion can evolve if one always follows the principle of conditionalization, but there can be no radical revisions or clean breaks with past opinion. It seems that people do sometimes change

---

[1] I want to thank Prasanta S. Bandyopadhyay for helpful discussion.

[2] I assume here that the agent's degrees of confidence or degrees of belief satisfy the probability axioms.

their opinions in ways other than by conditionalization on what they have learned, and moreover there are times when they are surely right to do so. While it is enormously difficult to characterize what makes a judgment a good one, it is obvious that some judgments are better than others and there is typically agreement that certain judgments are good and others poor. Suppose that my probability for being rich tomorrow conditional on my spouse buying a ticket in the state lottery is high and that I learn subsequently that my spouse has indeed purchased a lottery ticket. If I update by conditionalization, then I should attach a high probability to being rich tomorrow. But, it seems clear that since I am not rich now, and have no prospects for great wealth tomorrow, besides the lottery, that I should not attach a high probability to being wealthy tomorrow. This would be a foolish judgment, given my circumstances, and so was my prior judgment. If a person's judgments are inadequate, then it seems clear that she should not base her future opinion on her current opinion in accordance with the principle of conditionalization, but rather should, if feasible, revise her probabilities so as to reflect good judgment.[3]

Such examples appear to show that it is not always reasonable to change beliefs by conditionalization and hence that the principle of conditionalization, as characterized above, is false. Thus, it is not merely "unrealistic," to try to show that the principle of conditionalization is a requirement of rationality, as Paul Teller has noted in "Conditionalization and Observation"(Teller 1973, p. 220), but misguided. Still, the more modest goal, addressed by Teller, of showing that "under certain well specified conditions, only changes by conditionalization are reasonable,"[4] might still be successfully pursued.

The most common way of trying to establish that, under certain conditions, one ought to change beliefs by conditionalization appeals to the so-called Dutch Strategy argument.[5] A Dutch Strategy, like a Dutch Book, consists of a series of bets, which guarantees the bettor a net loss; however, in the case of a Dutch Strategy the bets are made over time, whereas a Dutch Book involves bets made a single time. If an agent's beliefs change, after learning E, by a rule other than conditionalization, then she is susceptible to a Dutch Strategy, in that a bookie, who knows her degrees of confidence and her rule for updating, can devise a series

---

[3]There are other cases where it would be better not to conditionalize. Indeed, assuming the Bayesian principle that one should act so as to maximize expected utility, then whenever the circumstances are such that the greatest expected utility involves violating conditionalization, it will be rational to do so. One way of producing such a case is just to attach a sufficiently large prize to a violation of conditionalization. Patrick Maher (1993) has given several examples of this sort.

[4]Ibid.

[5]The argument was constructed by David Lewis and is reported in Teller (1973).

of bets to be placed at different times, each of which should appear fair to the agent at the time offered, but which together guarantee her a net loss.[6]

Many questions have been raised about the force of the Dutch Strategy argument to show that an agent's degrees of belief should change by conditionalization, as well as about the basic Dutch Book argument to show that an agent's degrees of confidence should satisfy the probability axioms (i.e. that they are coherent). The basic Dutch Book argument has been thought to show that a rational agent's degrees of confidence must satisfy the probability axioms, because failure to satisfy the axioms leaves one susceptible to a Dutch Book, which in turn would mean a sure loss (Jackson and Pargetter 1976). However, many reasons have been given for thinking that it is not necessarily irrational to violate the probability axioms. First, leaving oneself open to a Dutch Book, by having degrees of confidence that do not satisfy the probability axioms does not insure victimization by a clever bookie. Agents are unlikely to encounter a bookie who has the knowledge and desire to take advantage of them and, in any case, an agent can avoid having a Dutch Book made against them by simply refusing to bet. Second, it is not always irrational to guarantee that one will suffer a net monetary loss, since such a loss might involve non-monetary gains.[7] Similar objections would clearly apply to the Dutch Strategy argument for conditionalization. It appears that if the Dutch Book and Dutch Strategy arguments are supposed to work by

---

[6]If the agent plans to change probabilities if E is learned such that $p'(A) < p'(A|E)$, then the bookie can guarantee the agent a net loss by selling the agent the following bets:

1	if $A\&E$	for the price $p(A\&E)$
0	otherwise	

$p(A|E)$	if $-E$	for the price $p(A|E)p(-E)$
0	otherwise	

$p(A|E) - p'(A)$	if E	for the price $[p(A|E) - p'(A)]p(E)$
0	otherwise	

If E is false, the agent has a net loss of $[p(A|E) - p'(A)]p(E)$. If E is true, then the bookie buys back the bet

1	if A	for the price $p'(A)$
0	otherwise	

Thus, if E is true the agent also suffers a net loss of $[p(A|E) - p'(A)]p(E)$. If the agent's rule is such that $p'(A) > p(A|E)$, the bookie simply reverses the direction of these bets to make a Dutch Book.

[7]For detailed criticisms along these lines see Adams and Rosenkrantz (1980); Kennedy and Chihara (1979); and Howson and Urbach (1993).

showing that violation of the probability axioms leads to undesirable consequences, and hence such a violation is irrational, then the arguments are not very convincing.

Recently, it has been suggested by various philosophers that the Dutch Book argument is misunderstood if it is thought to force compliance with the probability axioms as a means of avoiding a monetary loss (Skyrms 1987; Howson and Urbach 1993). Instead, they suggest that the possibility of making a Dutch Book against a set of betting quotients reveals that the beliefs represented by those betting quotients suffer from a form of inconsistency,[8] Indeed, it is the inconsistency that makes the Dutch Book possible. This is seen quite easily in the case where an agent violates the axiom of the probability calculus that requires that $p(A \vee B) = p(A) + p(B)$, when A and B are mutually exclusive. Since a bet on A and a bet on B is equivalent to a bet on $(A \vee B)$, having a degree of belief for $(A \vee B)$, which differs from the sum of that for A and for B is equivalent to having two different evaluations of $(A \vee B)$, and it is the difference (inconsistency) that the clever bookie exploits. Similarly, in the Dutch Strategy argument for conditionalization, the bookie takes advantage of the difference between the agent's present conditional probability of A given E and the probability she plans to attach to A upon learning E.

Once it is seen that Dutch Book vulnerability arises from a kind of inconsistency, the problem is to determine the kind of defect signaled by that vulnerability. Such vulnerability is not always irrational, since there can be goods of various kinds associated with exhibiting an inconsistency of any sort. Nevertheless, to have in effect two different evaluations for the proposition A or B, or to have less than full confidence in a tautology, does involve a kind of epistemic defect. The observation that violation of the probability axioms involves a kind of inconsistency would appear to explain why Dutch Strategy vulnerability involves a defect. However, it has recently been argued by David Christensen (1991) that while Dutch Book vulnerability always stems from some sort of inconsistency, the inconsistency does not always constitute an epistemic defect. In particular, Christensen claims that the inconsistency involved in the Dutch Strategy

---

[8]Indeed, this interpretation of the Dutch Book argument certainly seems to reflect the views of its originator F.P. Ramsey.

> Any definite set of degrees of belief which broke them would be inconsistent in the sense that it violated the laws of preference between options, . . . If anyone's mental condition violated these laws, his choice would depend on the precise form in which the options were offered him, which would be absurd. He could have book made against him by a cunning bettor and would then stand to lose in any event. (Ramsey 1926/1990, p. 78)

argument for conditionalization is diachronic, and that epistemic rationality requires synchronic, but not diachronic, consistency.[9]

A closer look at the details of the Dutch Strategy argument for conditionalization suggests that vulnerability to a Dutch Strategy does involve a form of synchronic inconsistency, despite the fact that conditionalization is a principle of belief change. Dutch Strategy vulnerability does not result from the inconsistency involved in merely holding one set of beliefs at one time and a different set at a later time, or even by changing beliefs in a way that is not given by conditionalizing on what is learned. In order for a bookie to take advantage of an agent who violates conditionalization, the bookie must know the agent's rule for changing his beliefs in advance of any change of belief. The Dutch Strategy argument for conditionalization thus assumes that the agent is committed to changing his degree of belief in a proposition A upon learning E, to a value which differs from his current probability of A given E.[10] The inconsistency that the Dutch Strategy turns on is synchronic, since it involves holding a certain set of prior beliefs and *at the same time* holding that were some new evidence E obtained one's beliefs would change in a particular way which does not go by conditionalization on E.

To have a Dutch Strategy against a non-conditionalizing agent requires knowing the agent's deviant updating rule. As van Fraassen has observed, this means that it is always possible to avoid a Dutch Strategy, even if, in the end, one does not change beliefs by conditionalization, by not announcing in advance the rule one will follow (van Fraassen 1989). If indeed Dutch Strategy vulnerability is to be avoided on epistemic grounds, as opposed to pragmatic ones, then it is not enough that one simply refrain from *announcing* a deviant rule, one must not *plan* in advance to violate conditionalization, since it would be having such a plan that produces the inconsistency. There is nothing in the Dutch Strategy argument itself that entails that one plan in advance how one will change beliefs upon learning that a certain proposition is true. Hence the Dutch Strategy argument alone does not support what I have called the principle of conditionalization, which *requires* changing beliefs by conditionalization.

If the Dutch Strategy argument does not show that we must conditionalize, what if anything does it show? Van Fraassen has written that the argument shows that the only acceptable rule of belief change that applies when a proposition is learned is conditionalization, although van Fraassen

---

[9]For discussion and criticism of Christensen's analysis, see Vineberg (1997).

[10]If the agent were not committed to changing his beliefs by some deviant rule, the bookie could not be certain of profit. However, the commitment could be less than full. In this case the bookie would need to make a side bet against the possibility that the agent reneges in order to insure a net profit.

does not believe that rationality requires rule following.[11] Why though is conditionalization on E the only rule for updating when E is learned? It could be said that adopting an alternative rule would leave one exposed to a potential loss, but this would just be to invoke the discredited interpretation of Dutch Book arguments, as turning on the pragmatic consequences of Dutch Book vulnerability. An alternative answer is that planning to change beliefs in a way that violates the principle of conditionalization involves having inconsistent degrees of confidence. When an agent adopts a deviant rule, the bookie's guarantee of a net gain is fixed prior to the change by the fact that there is a set of conditional odds on E associated with the deviant rule that differs from the agent's current set of conditional odds on E. It seems though that endorsing such a deviant rule does not strictly entail having inconsistent beliefs. The problem is that it isn't clear that endorsing such a rule now entails actually having the degrees of confidence that would be had if one's beliefs were adjusted in accordance with that rule. However, endorsing a rule which involves changing beliefs upon learning E in a way that does not go by conditionalization on E would seem to involve a simultaneous commitment to incompatible odds. In any case, I suggest that efforts to identify violations of conditionalization that lead to Dutch Strategy vulnerability with inconsistency ignore the character of conditionalization as a rule. What ultimately makes the Dutch Strategy argument possible is the fact that when E is learned the only rule which preserves conditional odds on E is the conditionalization rule. This suggests that the appropriate way of understanding conditionalization as a part of the logic of belief, along with the probability axioms as a consistency constraint, is as a rule of inference. Thus interpreted, we should read the rule as stating that from one's prior probability for a hypothesis H conditional on E and given a new probability for E of one, it follows that the new probability for H is the prior probability of H given E. As with the standard logical inference rules, it is a rule of permission, rather than a requirement of rationality. Just as Modus Ponens does not compel one who accepts 'if A then B' and learns A to infer B, rather than giving up 'if A then B,' so there is no rational requirement that the hypothesis H be updated by conditionalizing on E, when E is learned, as opposed to giving up the prior conditional probability of H given E. All that is required is that if the odds conditional on E are accepted as fixed, then probabilities must shift by conditionalization upon learning E.

---

[11] More precisely, van Fraassen's view is that conditionalization is the only updating rule in the cases where it applies, that is, when the probability of some proposition E shifts to one. According to van Fraassen (1989), there can indeed be other rules that apply in other circumstances.

2. Confirmation

I have been arguing that the Dutch Strategy argument does not support the claim that when a proposition E is learned that rational beliefs must shift by conditionalization on E, but rather points to conditionalization's status as a kind of inference rule.[12] In the remainder of the paper, I want to consider the consequences of this point for Bayesian confirmation theory.

According to the Bayesian theory of confirmation evidence E confirms a hypothesis H iff $p(H|E) > p(H)$. If the *principle* of conditionalization is also assumed, then evidence E that confirms a theory, is evidence which, if obtained, would raise the probability of the theory. Suppose though that the principle of conditionalization is not always followed when new evidence is obtained and that sometimes instead probability judgments of propositions conditional on the evidence are given up when the evidence comes in. How does this affect the Bayesian treatment of confirmation?

Suppose that an agent has probabilities at t_1 such that $p_1(H|E) > p_1(H)$. Prior to learning E, the agent should say that, by his current opinion, learning E would confirm H. Assume that at time t_2 E is learned, but the agent actually lowers his probability for H. Here he should say that, although prior to learning E he thought that E would confirm H, upon learning E he did not take it as confirming H and that he no longer endorses his previous view that learning E would confirm H. While such situations will occur, it is probably fair to say that they are not typical of testing situations in science, so that normally when E is taken, prior to learning that E is true, as having positive evidential bearing on H, learning E will in fact raise the probability of H.[13]

Although Bayesians generally interpret probabilities as personal judgments, which may differ from person to person, science strives for consensus. It is not enough, within scientific communities, that some individual scientists have surveyed the evidence and have pronounced a hypothesis highly confirmed, since others may have quite different views. In cases

---

[12] There are, of course, other arguments besides the Dutch Strategy argument for the principle of conditionalization. One such argument is presented by Teller (1973). In additional, various attempts have been made to show that shifts in belief by conditionalization are, in a certain sense, minimal shifts (Diaconis and Zabell 1982). However, neither of these arguments establish that one must always update by conditionalization in those cases where the conditionalization rule applies.

[13] One place where this arises is during scientific revolutions. There probabilities will not change by conditionalization, although a strict Bayesian might argue that this is not a violation of the principle that *rational* changes of belief go by conditionalization. Whether, or how often, such violations of conditionalization occur within what Kurtz called normal science is difficult to answer, in part because it is so hard to specify the bounds of normal science.

where there is initial disagreement, Bayesians point proudly to the so-called merger of opinion results, which they claim make it likely that, in the long run, such initial diverse opinions will converge with accumulating evidence (see Edwards, Lindman *et al.* 1963). But, these merger of opinion results presuppose updating by conditionalization on the evidence. If there is no requirement that agents conditionalize, then the merger of opinion theorems provide no assurance of convergence.[14] Still, assuming that violations of conditionalization are rare, it may be that in typical scientific contexts, convergence will occur through conditionalization on empirical evidence (although, see Earman (1992)). However, even assuming strict adherence to the principle of conditionalization, there may be no convergence.

In order to illustrate the limits of conditionalization in reaching consensus it will be helpful to examine Jon Dorling's (1992) recent attempt to show that the principle of Bayesian conditionalization can resolve realist/antirealist debates. As Dorling sees these debates, the difference between realists and antirealists (which Dorling calls positivists) really just amounts to a difference in the level of confidence they have in the evidential support provided by empirical evidence for theories involving unobservables. Moreover Dorling claims that such differences can (at least sometimes) be bridged through conditionalizing on further observations. While I think there is little plausibility to the idea that all philosophical debates between realists and antirealists amount to just differing levels of confidence,[15] which can potentially be resolved through the accumulation of new evidence, there is at least more plausibility to the idea that realist/antirealist disputes *within* scientific disciplines can be so resolved.

To show that conditionalization can resolve disputes between realists and antirealists, Dorling provides a detailed analysis of the historical debate of the nineteenth century over the atomic hypothesis. In this case

---

[14]For a thorough discussion of the limitations of the merger of opinion theorems, see Earman (1992).

[15]There two different, though not unrelated, debates that rage between realists and antirealists. One concerns the extent of knowledge and the other concerns meaning. While Dorling claims that "Bishop Berkeley is simply less confident than the realist that his re-entering-the-study-like experiences will always be followed by as-if-perceiving-his desk appearances" (1992, p. 377), this fails to characterize the nature of difference between Berkeley and the realist. Berkeley and the realist think there are tables and chairs in the study, but differ over what it means to make this claim. It is doubtful that Berkeley really is less confident than the realist in the way Dorling claims, since for Berkeley it is God who guarantees that his re-entering-the-study-like experiences will always be followed by as-if-perceiving-his desk appearances. While differences over the extent of human knowledge may, in the end, be no more than differences in confidence levels, it is implausible that the realist/antirealist debates over meaning can be reduced to differences in confidence.

there are two theories to consider: the realist's theory T_r, which should be identified with an early nineteenth century version of the atomic theory and the observational consequences of the theory T_p, which consist of the laws of constant and multiple proportions. Since T_p consists of the observational consequences of T_r, T_r entails T_p, but T_p does not entail T_r. As Dorling understands the difference between the realist and antirealist, the realist attaches a higher probability to T_r given T_p than the antirealist. However, as evidence for T_p accumulates the realist's confidence in T_r will increase, but so too may the positivist's confidence in T_r.

As Dorling shows, the positivist can actually become more confident than not that the realist theory is true. In the historical case involving the debate between the (realist) atomic theory and the (positivist) law of constant and multiple proportions, it is reasonable to suppose that the observational evidence increased the probability of T_p. To see how this in turn affects the probability of T_r through Bayesian conditionalization, let p_r be the probability function of the realist and let p_p be the probability function of the antirealist or positivist. If we suppose (with Dorling) that the observational evidence drives the probability of T_p to one for both the realist and antirealist, then we may use the following instances of Bayes' Theorem to calculate the effect of this change on the probability of the realist hypothesis T_r:

(1) $p_r(T_r|T_p) = p_r(T_p|T_r)p_r(T_r)/p_r(T_p)$

(2) $p_p(T_r|T_p) = p_p(T_p|T_r)p_p(T_r)/p_p(T_p)$.

If we assume here that updating goes by conditionalization, then as the probability of T_p approaches one, $p_r(T_r)$ and $p_p(T_r)$ approach $p_r(T_r)/p_r(T_p)$ and $p_p(T_r)/p_p(T_p)$ respectively, since T_r entails T_p and so $p_r(T_p|T_r)$ and $p_p(T_p|T_r)$ both equal one. Dorling considers a particular case in which the realist has an initial probability in T_r of .6, but the antirealist only assigns a probability of .2 to T_r. Given these values we can then use the probability axioms to calculate $p_p(T_p)$ and $p_r(T_p)$. By the theorem of total probability we have:

(3) $p_r(T_p) = p_r(T_p|T_r)p_r(T_r) + p_r(T_p|-T_r)p_r(-T_r)$

(4) $p_p(T_p) = p_p(T_p|T_r)p_p(T_r) + p_p(T_p|-T_r)p_p(-T_r)$.

Dorling assumes that in this case the realist and antirealist agree initially on the probability that the law of constant and multiple proportions is true given that the atomic theory is false, that is that $p_r(T_p|-T_r) = p_p(T_p|-T_r)$. This need not be the case, but I will not dispute the assumption here. As an example, Dorling assumes that $p_r(T_p|-T_r) = p_p(T_p|1-T_r) = .2$. Substituting these values into (3) and (4) we have

(5) $p_r(T_p) = 1 \times .6 + .2 \times .4 = .68$

(6) $p_p(T_p) = 1 \times .2 + .2 \times .8 = .36.$

Using these values and substituting into (1) and (2), we have

$$p_r(T_r|T_p) = .6/.68 \approx .9$$

$$p_p(T_r|T_p) = .2/.36 \approx .6.$$

Apparently, Dorling regards holding a probability for T_r over .5, as endorsing realism, since he describes this situation as one in which evidence for T_p converts the positivist into a realist!

However, as Dorling himself observes, if the positivist's initial probability for T_r is low enough, he will never convert to realism through the application of Bayes' Law. In the example above, this would occur if rather than an initial probability of .2 for T_r, the antirealist assigns an initial probability of .1 to the realist hypothesis. But, if that is the case, then instead of adherence to the principle of conditionalization leading to a convergence of beliefs between the realist and antirealist, strict adherence to the Bayesian Principle of conditionalization will actually block resolution of the realist/antirealist debate. Even if the antirealist's probability for T_r does reach .5 it is doubtful that this really counts as a conversion to realism or represents any genuine merger of opinion, since the realist's probability for T_p will typically be much higher. What the previous example shows is that in some cases even this minimal sort of agreement may fail to be reached through conditionalization.

The fact that conditionalization on observational evidence does not always lead to a resolution of such debates fits well with the empiricist's position that no empirical observation would ever require acceptance of scientific realism. For instance, van Fraassen argues that we should regard observational evidence for a theory as evidence that the theory is empirically adequate, i.e. that the observational predictions of the theory are true, but not that the non-observational consequences are true.[16] Such an empiricist maintains that accumulation of observational evidence for T_p would not involve conversion to realism. Yet, as least on the Bayesian analysis of evidence, observational data does raise the probability of realist theories. So a diehard Bayesian antirealist would need to maintain that the initial probability of a given realist hypothesis is indeed very low, so that subsequent observational evidence would not raise the probability high enough for conversion to realism.

---

[16]This is slightly inaccurate since van Fraassen (1980) identifies scientific theories with models rather than collections of sentences.

I suggest that the standard considerations in favor of an antirealist stance towards theories shows why the antirealist should indeed attach very low initial probabilities to any theory that involves unobservables. The driving force behind empiricism in the philosophy of science is that for any theory T_r that involves claims about unobservables, there will be another empirically equivalent theory $T_{r'}$ that is just like the first with respect to its claims about observables, but which makes different claims about unobservables.[17] The antirealist claims that there is no observational evidence which could distinguish T_r from $T_{r'}$. Moreover, there is not just a single theory $T_{r'}$ that the antirealist maintains is evidentially indistinguishable from T_r, but infinitely many such theories. Given a theory T_r, scientists will generally have little need to formulate empirically equivalent alternatives, and so typically there will be no fully articulated theories that are empirically equivalent to T_r that have been formulated as rivals to T_r. Nevertheless, the empiricist takes such, as yet unformulated, rivals seriously. Let **T** be the class of theories that are empirically equivalent to T_r. The antirealist's probability for a specific theory T_r will be very small in comparison to the probability that one of the theories in **T** is true. Scientists typically do not take various empirically equivalent theories as serious alternatives, and so may be thought of as assigning initial probabilities that do allow for resolution of realist/antirealist debates in accordance with Bayesian conditionalization.

It follows from Bayes' Theorem $[p_r(T_r|T_p) = p_r(T_p|T_r)p_r(T_r)/p_r(T_p)]$ that the individual assessments of the probability of T_r and T_p are connected with the evidential bearing of T_p on T_r. These assessments depend, in turn, on various epistemic assumptions on which realists and antirealist typically disagree, including assumptions about what constitutes the relevant alternatives to a theory. Understanding the conditionalization rule as a kind of inference rule is helpful in seeing why steadfastly following the principle of conditionalization cannot generally resolve realist/antirealist disputes. Inference rules cannot resolve fundamental differences in starting assumptions, but at best show us the consequences of what we already accept, or what these assumptions entail in conjunction with new information. In cases where the theory in question outstrips the potential evidence for it, conditionalization on that evidence won't necessarily lead to significant agreement. In such cases, reaching a consensus involves violating the principle of conditionalization.

REFERENCES

Adams, E. W. and R. D. Rosenkrantz. (1980). Applying the Jeffrey Decision Model

---

[17]For a critical discussion of this motivation for antirealism, see Boyd (1984).

to Rational Betting and Information Acquisition. *Theory and Decision* **12**, 1–20.

Boyd, R. (1984). *The Current Status of Scientific Realism*. *Scientific Realism*. Berkeley: University of California Press.

Christensen, D. (1991). Clever Bookies and Coherent Beliefs. *The Philosophical Review* **100**, 229–247.

Diaconis, P. and S. L. Zabell. (1982). Updating Subjective Probability. *Journal of the American Statistical Association* **77**, 822–30.

Dorling, J. (1992). Bayesian Conditionalization Resolves Positivist/Realist Disputes. *The Journal of Philosophy* **89**, 362–82.

Earman, J. (1992). *Bayes or Bust?* Cambridge: MIT Press.

Edwards, W., H. Lindman and L. J. Savage. (1963). Bayesian Statistical Inference for Psychological Research. *Psychological Review* **70**, 193–242.

Howson, C. and P. Urbach. (1993). *Scientific Reasoning: The Bayesian Approach*. LaSalle, Illinois: Open Court.

Jackson, F. and R. Pargetter. (1976). A Modified Dutch Book Argument. *Philosophical Studies* **29**, 403–407.

Kennedy, R. and C. Chihara. (1979). The Dutch Book Argument: Its Logical Flaws, Its Subjective Sources. *Philosophical Studies* **36**, 19–33.

Kuhn, T. S. (1962). *The Structure of Scientific Revolutions*. Chicago: The University of Chicago Press.

Maher, P. (1993). *Betting on Theories*. Cambridge: Cambridge University Press.

Ramsey, P. F. (1926/1990). Truth and Probability. In D. H. Mellor (Ed.), *P. F. Ramsey Philosophical Papers*. Cambridge: Cambridge Univ. Press. pp. 52–95.

Skyrms, B. (1987). Coherence. In N. Rescher (Ed.), *Scientific Inquiry in Philosophical Perspective*. Pittsburgh: University of Pittsburgh Press. pp. 225-242.

Teller, P. (1973). Conditionalization and Observation. *Synthese* **26**, 218–258.

van Fraassen, B. C. (1980). *The Scientific Image*. Oxford: Oxford University Press.

van Fraassen, B. C. (1989). *Laws and Symmetry*. New York: Oxford University Press.

Vineberg, S. (1997). Dutch Books, Dutch Strategies and What They Show About Rationality. *Philosophical Studies* **86**, 185–201.

Susan Vineberg
Department of Philosophy
Wayne State University
Detroit, Michigan 48202
e-mail: susan.vineberg@wayne.edu

Commentary by Piers Rawling

Vineberg suggests that we should treat the principle of updating by conditionalization (*C*) as an inference rule, and then goes on to explore the ramifications of this for Bayesian confirmation theory. While I agree with her conclusion that updating by conditionalization "cannot generally resolve realist/anti-realist disputes" (p. 87), I shall dispute both her claim that *C* can be treated as an inference rule and her claim that viewing *C* in this light aids in our understanding of why updating by conditionalization does not necessarily result in convergence of opinion.

One aspect of Vineberg's claim that C should be treated as an inference rule is the analogy she endeavors to draw between C and Modus Ponens (MP), a key aspect of which, for her purposes, is the claim that both C and MP rule certain sets inconsistent. MP rules the following set inconsistent: $\{A, A \longrightarrow B, \sim B\}$. Thus MP dictates that at least one of the three propositions must be abandoned, but it does not dictate which. The analogous notion vis-a-vis C might be that the following set of probabilistic judgments is inconsistent: $\{p(H|E) = p, p(E) = 1, p(H) = q \neq p\}$, and one of the three must be modified, but C does not dictate which. In the case of confirmation, initial opinions might differ to such an extent that convergence will not necessarily be reached by conditionalizing on the evidence: consensus might require modification of conditional probabilities.

The analogy between probabilistic and logical inconsistency apparently runs fairly deep. Vineberg considers the case of an agent who believes the propositions A and B to be mutually exclusive, and yet does not equate the sum of $p(A)$ and $p(B)$ to $p(A \vee B)$. Suppose $p(A) = 0.4$, $p(B) = 0.3$, and A and B are mutually exclusive. Then a fair bet on A is:

G1   win \$6 if A, lose \$4 if $\sim A$;

and a fair bet on B is:

G2   win \$7 if B, lose \$3 if $\sim B$.

On the proviso that A and B are mutually exclusive, taking G1 and G2 simultaneously is equivalent to taking the single bet:

G3   win \$3 if $A \vee B$, lose \$7 if $\sim A \& \sim B$

One can arrive at this equivalence either by noting that $0.3 + 0.4 = 0.7$ and invoking:

$$\text{If } \sim (A\&B) \text{ then } p(A) + p(B) = p(A \vee B);$$
$$(\{\sim (A\&B), p(A) + p(B) \neq p(A \vee B)\} \text{ is inconsistent})$$

or by invoking:

$$\text{If } \sim (A\&B) \text{ then } [((A\& \sim B) \vee (\sim A\&B)) \equiv (A \vee B)]$$
$$(\{\sim (A\&B), \sim [((A\& \sim B) \vee (\sim A\&B)) \equiv (A \vee B)]\} \text{ is inconsistent}).$$

In the case of an agent who fails to see that:

$$\text{If } \sim (A\&B) \text{ then taking G1 and G2 is equivalent to taking G3,}$$

probabilistic inconsistency parallels logical inconsistency.

What is the logical parallel to the inconsistency of $\{p(H|E) = p, p(E) = 1, p(H) \neq p\}$? One possibility is to note that if $p(H\&E) = p$ and $p(E) = 1$ then $p(H|E) = p$, and hence the inconsistency of $\{p(H|E) = p, p(E) = 1, p(H) \neq p\}$ entails the inconsistency of $\{p(H\&E) = p, p(E) = 1, p(H) \neq p\}$. The latter inconsistency is evinced by an agent who fails to see that the combination of:

G4   win \$1 if $(H\& \sim E)$, lose \$0 if $\sim (H\& \sim E)$

and:

G5   win \$$(1 - p)$ if $(H\&E)$, lose \$$p$ if $\sim (H\&E)$

is equivalent to:

G6   win \$$(1 - p)$ if H, lose \$$p$ if $\sim H$.

Noting the equivalence of G6 to the combination of G4 and G5 is tantamount to noting the equivalence of H to

$$[((H\& \sim E)\& \sim (H\&E)) \vee (\sim (H\& \sim E)\&(H\&E))]$$

and $\sim H$ to

$$[\sim (H\& \sim E)\& \sim (H\&E)]$$

— i.e., tantamount to noting the inconsistency of the sets

$$\{\sim [[((H\& \sim E)\& \sim (H\&E)) \vee (\sim (H\& \sim E)\&(H\&E))] \equiv H]\}$$

and

$$\{\sim [\sim H \equiv [\sim (H\& \sim E)\& \sim (H\&E)]]\}.$$

So, if C is read as asserting the inconsistency of $\{p(H|E) = p, p(E) = 1, p(H) \neq p\}$, then the agent whose degrees of belief do not conform to C can be seen as falling prey to a synchronic logical inconsistency. There is a key difficulty here however: in the discussion of C, I have ignored the distinction between prior and posterior probabilities. The inconsistency of $\{p(H|E) = p, p(E) = 1, p(H) = q \neq p\}$ follows from Kolmogorov's axioms and his definition of conditional probability. C, however, is meant to be a distinct principle. One possibility, which Vineberg dismisses, is that violations of C are diachronically inconsistent. Vineberg claims, rather, that such violations are synchronically inconsistent because violating C "involves holding a certain set of prior beliefs and *at the same time* holding that were some new evidence E obtained one's beliefs would change in a particular way which does not go by conditionalization on E" (p. 78). One difficulty with this claim is that it runs counter to certain views as to the very nature of conditional probability: Savage (1972, p. 44), for example, claims that "$p(C/B)$ can be regarded as the probability the person would assign to C after he had observed that B obtains."

However, let us suppose for the moment that we can make sense of the notion of conditional probability independently of a commitment to C, and explore Vineberg's claim that "there is no rational requirement that the hypothesis H be updated by conditionalizing on E, when E is learned, as opposed to giving up the prior conditional probability of H given E" (p. 82). Perhaps this better captures Vineberg's treatment of C as an inference rule than the thought that $\{p(H\&E) = p, p(E) = 1, p(H) \neq p\}$ is inconsistent. However, as it stands, the claim is (almost) trivially true — as Vineberg notes: "[t]he principle of conditionalization is generally stated as requiring that an agent's new probability for A after learning E, *and nothing more*, should be equal to her old probability of A given E" (emphasis mine, p. 77). Thus Vineberg's claim should be modified to read:

> There is no rational requirement that the hypothesis H be updated by conditionalizing on E, when *only* E is learned, as opposed to giving up the prior conditional probability of H given E.

But this is, I think, simply false. Suppose an agent holds $p(H|E) = p$ at t and $p'(H) = q \neq p$ at t' (later than t) having learned only E in the interim. Of course, the conjunction of $p'(H) = q$ and $p'(E) = 1$ entails that $p'(H|E) = q$. Thus the agent has modified her degree of belief in $[H$ given $E]$ having learned only E. The issue is: how can this be, except on a whim? Surely, for such a modification to be *rational*, the agent must have learned something besides E — perhaps that she was irrational or mistaken in her earlier judgment, for example. Suppose otherwise. Then we would want to say, for example, something about what *degree* of change is rational in any case in which we claim that it is rational to change $p(H|E)$ having learned only E. And I do not see how to do this.

I have so far argued against two of Vineberg's claims. First, if we treat C as asserting the inconsistency of $\{p(H|E) = p, p(E) = 1, p(H) \neq p\}$, we elide the distinction between prior and posterior probabilities. Second, I cannot make sense of Vineberg's claim that a rational agent can hold $p(H|E) = p$ at t and $p'(H) = q \neq p$ at t' (later than t) having learned only E in the interim. Suppose, however, that we could make sense of the claim that we see C anew if we view it as an inference rule. I turn now to dispute Vineberg's claim that so understanding C "is helpful in seeing why steadfastly following the principle of conditionalization cannot generally resolve realist/antirealist disputes" (p. 87).

The reason Vineberg cites for lack of convergence here seems to be the entirely standard one of differing priors. Suppose both sides agree that $p(E|H) = 1$ and $p(E| \sim H) = 0.2$, but on one side $p(H) = 0.1$ whereas on the other $p(H) = 0.9$. It follows from Bayes' Theorem that in the former case $p(H|E) = 0.98$, and in the latter $p(H|E) = 0.36$. Supposing only E is learned by both sides, and applying C, we have respective values for $p'(H)$ of 0.98 and 0.36 — far from convergence. The notion that C is an inference rule has not entered. Certainly, convergence here can be reached only if $p'(H)$ differs from $p(H|E)$ for one side or the other or both. However, such convergence can only be rationally arrived at if one side or the other or both learn something in addition to the truth of E — in which case, pace Vineberg, consensus is reached without violating C. Admittedly, if something is learned in addition to E, it might be argued that whilst C is not violated, it is not followed either. But it is easily shown that not *all* learning can be by conditionalization (see Jeffrey 1983, Ch. 11).

I conclude not only that C cannot be regarded as an inference rule in the way that Vineberg suggests, but also that viewing C in standard fashion does not vitiate its role in explaining lack of convergence. I suspect that part of the problem with Vineberg's approach is that she ignores some fundamental disanalogies between C and MP. First, adding a proposition to the set $\{A, A \longrightarrow B, \sim B\}$ cannot resolve its inconsistency; however adding that some proposition (F, say) has been learned in addition to E can justify an inequality between $p'(H)$ and $p(H/E)$, and the learning of some such proposition F is necessary for such justification. Second, nothing is necessarily implicitly rejected by an agent for whom $p'(H) \neq p(H|E)$: $p(H|E)$ was what it was, $p'(H)$ (which equals $p'(H|E)$ if $p'(E) = 1$) is what it is, and both valuations might well be justified.

REFERENCES

Savage, L. (1972). *Foundations of Statistics*, 2nd ed. New York: Dover.
Jeffrey, R. (1983). *The Logic of Decision*, 2nd ed. Chicago: University of Chicago Press.

Piers Rawling
Department of Philosophy
University of Missouri, St. Louis
St. Louis MO 63121-4499
e-mail: sjprawl@umslvma.umsl.edu

Vineberg's Reply to Rawling

In my paper, I distinguished the diachronic Principle of Conditionalization (*PC*) from what may be called the Rule of Conditionalization (*RC*). *PC* says that when *e* and nothing stronger is learned, an agent should always update her degrees of belief by conditionalization on *e*. *RC*, on the other hand, makes no claim whatsoever about how rational agents ought to change their degrees of belief. Rather *RC*, which is analogous in important ways to Modus Ponens,[18] says that where the new probability for *e* is set equal to one, and where the probability for *A* conditional on *e* remains fixed, the new probability for *A* goes by conditionalization on *e*. That is, if $p'(e) = 1$ and $p'(A|e) = p(A|e)$, then $p'(A) = p(A|e)$.

The so-called Dutch Strategy argument for conditionalization appears to have been taken as supporting *PC*, but examination shows that it cannot do so. Rather, the Dutch Strategy vulnerability in question comes from holding at t_1 both that (1) $p(A|e) = r$, and (2) if $p'(e) = 1$, then $p'(A|e) = q \neq r$. Together these claims involve a commitment to two different sets of conditional odds for *A* on *e*, and hence result in Dutch Strategy vulnerability. The vulnerability comes from planning to violate what I have identified as *PC*, but one may violate *PC* without planning this in advance, and thereby avoid Dutch Book. The Dutch Strategy argument is connected with *RC*, in that as long as one is committed to a particular set of conditional odds for *A* on *e*, belief change upon learning *e* must go by conditionalization on *e*. However, this in no way requires that one must continue to accept that particular set of conditional odds on *e* after learning *e*, as required by *PC*.

So far, I have answered Rawling's claim that I read conditionalization in an implausible way that fails to respect the difference between prior and posterior probabilities. Rather, I have distinguished two claims involving conditionalization, each referring to prior and posterior probabilities. Rawling has three other objections, which I will briefly address. First, he disputes my claim that it can be rational to violate *PC*. However, a number of counterexamples to the principle have been given (Maher 1993; Vineberg 1991). Second, Rawling suggests that in discussing the Dutch Strategy argument, I depend on a questionable interpretation of conditional probability, in claiming that a person who plans to violate conditionalization has a current conditional probability for *A* given *e* that differs from the probability she would assign to *A* if in the future she learns that *e* is true. The complaint is that this characterization presupposes that a person's conditional probability for *A* given *e* may differ from

---

[18]Since my original paper was written, the analogy between *RC* and Modus Ponens has been developed independently by Howson (1997).

the probability that she would attach to A if she were to learn e. While Savage, among others, have taken these to be the same, if a person's conditional probability for A given e is associated with her fair betting quotient for a bet on A conditional on e, then this may in fact differ from what she thinks her fair betting quotient on A would be upon learning e, as recognized by Ramsey (1926/1990), among others. Finally, once it is acknowledged that even where PC applies, rationality does not dictate that it be strictly followed, the procedure of conditionalizing on new information can seen as a decision to apply RC. Where agents have priors that will not lead to convergence, regardless of the incoming evidence, the only possibility of eventual agreement will involve a decision not to apply RC in violation of PC.

REFERENCES

Howson, C. (1997). Logic and Probability. *British Journal for the Philosophy of Science* **48**, 517–531.

Maher, P. (1993). *Betting on Theories*. Cambridge Studies in Probability, Induction, and Decision Theory. Cambridge: Cambridge University Press.

Ramsey, P. F. (1926/1990). Truth and Probability. In D. H. Mellor (Ed.), *P. F. Ramsey Philosophical Papers*. Cambridge: Cambridge Univ. Press. pp. 52–95.

Vineberg, S. (1991). *Conditionalization and Rational Belief Change*. Ph.D. Dissertation, UC Berkeley.

*Poznań Studies in the Philosophy
of the Sciences and the Humanities
2000, vol. 71, pp. 95-111*

Deborah G. Mayo

SCIENCE, ERROR STATISTICS, AND
ARGUING FROM ERROR

ABSTRACT. I distinguish two main approaches to uncertain inference using probabilities: the evidential-relation (e.g., Bayesian) approach, and the
Neyman-Pearson error-statistical approach. By controlling the error probabilities of inference methods, I argue, the error-statistical approach offers powerful
tools for obtaining reliable experimental knowledge. I develop a framework
in which these tools, correctly understood, and suitably interpreted, form the
basis for a new and fruitful philosophy of experimental inference.

> *The two main attitudes held to-day towards the theory of probability both result
> from an attempt to define the probability number scale so that it may readily
> be put in gear with common processes of rational thought. For one school, the
> degree of confidence in a proposition, a quantity varying with the nature and
> extent of the evidence, provides the basic notion to which the numerical scale
> should be adjusted. The other school notes how in ordinary life a knowledge of
> the relative frequency of occurrence of a particular class of events in a series of
> repetitions has again and again an influence on conduct; it therefore suggests
> that it is through its link with relative frequency that a numerical probability
> measure has the most direct meaning for the human mind* (Pearson 1966, p.
> 228).

1. Introduction

The two main attitudes of which Pearson here speaks correspond to two
distinct views of the task of a theory of statistics: the first we may call the
evidential-relation view, and the second, the *error probability* or sampling
view. This difference corresponds to fundamental differences in the idea of
how probabilistic considerations enter in scientific inference. Evidential-
relationship, or E-R, approaches grew quite naturally from what was tra-
ditionally thought to be required by a "logic" of confirmation or induction.
Most commonly, such approaches seek quantitative measures of the bearing of evidence on hypotheses. What I call error statistical approaches, in
contrast, focus their attention on finding general methods or procedures
of testing with certain good properties.

A main way to contrast the E-R with the error statistical approach
is by means of their quantitative measures. The quantities in evidential-
relationship, E-R, approaches are probabilities or other measures (of support or credibility) assigned to hypotheses. In contrast, testing approaches
do not assign probabilities to hypotheses. The quantities and principles
in testing approaches refer only to properties of methods, e.g., of testing
or of estimation procedures. An example would be the probability that a
given procedure of testing would reject a null hypothesis erroneously —
an error probability.

In the E-R view, the task of a theory of statistics is to say, for given
evidence and hypotheses, how well evidence confirms or supports hypotheses (whether absolutely or comparatively). In this view — one embraced
by most philosophers of induction and statistics — the role of statistics
is that of furnishing a set of formal rules or "logic" relating given evidence to hypotheses. The dominant example of such an approach on the
contemporary philosophical scene is based on one or another Bayesian
measure of support or confirmation. With the Bayesian approach, what
we have learned about a hypothesis H from evidence e is measured by the
conditional probability of H given e using Bayes's Theorem. The corner-
stone of the Bayesian approach is the use of prior probability assignments
to hypotheses, generally interpreted as an agent's subjective degrees of
belief.

In contrast, the methods and models of classical and Neyman-Pearson
statistics (e.g., statistical significance tests, confidence interval methods)
are primary examples of error probability approaches. These eschew the
use of prior probabilities where these are not be based on objective frequencies. Probability enters instead as a way of characterizing the experimental or testing process itself; to express how reliably it discriminates
between alternative hypotheses and how well it facilitates learning from
error. These probabilistic properties of experimental procedures are *error
probabilities.*

Although methods and models from error statistics continue to dominate among experimental practitioners who use statistics, the Bayesian
way has increasingly been regarded as the model of choice among philosophers looking to statistical methodology to get at the logic of scientific
inference. Given the current climate in philosophy of science, it is surprising to find philosophers (still) declaring invalid a widely used set of
experimental methods, rather than trying to explain why scientists evidently (still) find them so useful. This has much less to do with any
sweeping criticisms of the standard approach than it does the fact that
the Bayesian view strikes a resonant cord with the logical-empiricist gene
inherited from early work in confirmation and induction. In any event,

I think the time is ripe to remedy this situation. A genuinely adequate philosophy of statistics will only emerge if it is not at odds with statistical practice in science.

My position is that the error statistical approach is at the heart of the widespread applications of statistical ideas in scientific inquiry, and that it offers a fruitful basis for a philosophy of experimental inference. However, there continues to be a good deal of dispute as to the adequacy of the error statistician's view of the role of probability in science. If probability is not being used to provide some sort of E-R measure, it is often asked, then what is the role of probability in scientific inference in this approach? My aim in this paper is to sketch an answer to this question. I shall consider, more specifically, the main threads that may be woven together to erect an error statistical philosophy of science.

2. Modeling Experimental Inquiry

The error statistical account utilizes and builds upon several methods and models from classical and Neyman-Pearson statistics, but it does so in ways that depart sufficiently from what is typically associated with these approaches as to warrant some new label. Nevertheless, I retain the chief feature of Neyman-Pearson methods – the centrality of error probabilities – hence the label "error-statistics." Moreover, what fundamentally distinguishes this approach from others is that in order to determine what inferences are licensed by data it is necessary to take into account the error probabilities of the experimental procedure.

By an error statistical approach to philosophy of science, I have in mind the various ways in which statistical methods based on error probabilities may be used in philosophy of science. A key role for probabilistic and statistical models in philosophy of science is to model experimental inference and inquiry.[1]

In contrast to an evidential-relationship view, rather than starting its work with evidence or data, our error-statistical approach includes

---

[1] Two other roles I identify are using statistical models to solve philosophical problems (e.g., about objectivity, underdetermination, progress) and using them to perform a methodological critique (e.g., a critique of the role of novel prediction, of the rule of varying data).

My reason for emphasizing the roles of statistics in philosophy of science is that it is with respect to these roles that the crucial differences between error statistical and Bayesian principles arise. In practice, in contrast, particular error statistical procedures often correspond to procedures Bayesian would countenance, albeit with differences in interpretation and justification. But in the uses of statistics in philosophy of science, these differences of interpretation and justification are paramount. For a fuller discussion, see Mayo (1996).

the task of arriving at data — a task that it recognizes as calling for its own inferences. A second point of contrast with other attempts that model scientific inference on statistical inference, whether Bayesian or non-Bayesian, is that it does not seek to equate the scientific inference with a direct application of some statistical inference scheme.

For example, to apply Neyman-Pearson statistics in philosophy of science, it is typically thought, requires viewing scientific inference as a matter of accepting or rejecting hypotheses according to whether outcomes fall in rejection regions of Neyman-Pearson tests. Finding that this distorts scientific inference, it is concluded that it is inappropriate to appeal to Neyman-Pearson statistics in erecting an account of inference in science. This conclusion, I have argued, is quite unwarranted because it overlooks the ways in which Neyman-Pearson methods, and standard statistics in general, are actually used in science. What I am calling the error statistical account, I believe, reflects these actual uses.

3. A Framework of Inquiry

To get at the use of these methods in science, I propose that experimental inference must be understood within a framework of inquiry. You cannot just throw some "evidence" at the error statistician and expect an informative answer to the question of what hypothesis it warrants. A framework of inquiry incorporates methods of experimental design, data generation, modeling and testing. For each experimental inquiry we can delineate three types of models: *models of primary scientific hypotheses*, *models of data*, and *models of experiment*. The following figure gives a schematic representation:

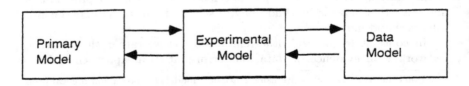

Figure 1. Models of Experimental Inquiry

A substantive scientific inquiry is to be broken down into one or more local or "topical"[2] hypotheses that make up the *primary questions* or *primary problems* of distinct inquiries. Typically, primary problems take the

[2]The term "topical hypotheses" is coined by Hacking (1992). Like topical creams, they are to be contrasted with deeply penetrating theories.

form of estimating quantities of a model or theory, or of testing hypothesized values of these quantities. These local problems often correspond to questions framed in terms of one or more standard or canonical errors: about parameter values, about causes, about accidental effects, and about assumptions involved in testing other errors. The *experimental models* serve as the key linkage models connecting the primary model to the data; links that require, not the raw data itself but appropriately *modeled data.*

Something like the above framework of models is hardly a new idea. It seems to me that a formal analog of the framework of models is found in standard statistical inference. I am not the first to think this. Patrick Suppes was. In his short but seminal paper, "Models of Data," Suppes remarks that:

> It is a fundamental contribution of modern mathematical statistics to have recognized the explicit need of a model in analyzing the significance of experimental data (Suppes 1969, p. 33).

In the error statistical account, formal statistical methods relate to experimental hypotheses, hypotheses framed in the experimental model of a given inquiry. Relating inferences about experimental hypotheses to primary scientific claims is, except in special cases, a distinct step. Yet a third step is called for to link raw data to data models — the real material of experimental inference. The indirect and piece-meal nature of our use of statistical methods, far from introducing an undesirable complexity into our approach, is what enables it to serve as an account of inference that is truly *ampliative.*

4. Severe Tests and Arguing From Error

Despite the complexity, there is an overall logic of experimental inference that emerges: Data e indicate the correctness of hypothesis H, to the extent that H passes a *severe test* with e. All of the associated inferential tasks are directed toward substantiating this piece of reasoning. Hypothesis H passes a severe test with e if (a) e fits H and (b) the test procedure had a high probability of producing a result that accords less well with H than e does, if H were false or incorrect.[3] To infer that H is indicated

---

[3]The severity requirement, more formally stated, is this:
Severity Requirement: Passing a test T with e) counts as a good test of or good evidence for H just to the extent that T is a *severe test* of H,
and the criterion of severity (SC) I suggest is this:
(1a): *Severity Criterion (SC)*: There is a very high probability that test procedure T would *not* yield such a passing result, if H is false.

by the data does not mean a high degree of probability is assigned to
H — no such probabilities are wanted or needed in the error statistical
account. That H is indicated by the data means that the data provide
good grounds for the correctness of H.

One can, if one likes, construe the correctness of H in terms of H being
reliable, provided one is careful in the latter's interpretation. Learning
hypothesis H is reliable means learning that what H says about certain
experimental results will often be close to the results actually produced —
that H will or would often succeed in specified experimental applications.
By means of statistical tests, we test whether this has in fact been learned.

Experimental learning in our account is learning about the (actual or
hypothetical) future performance of experimental processes — i.e., about
outcomes that would occur (with specified frequency) if certain experiments were carried out. This is *experimental knowledge*. Hypotheses asserting experimental knowledge may be inferred whether or not they are
part of any scientific theory. They have their own homes in experimental
models.

5. Learning From Error

The reasoning used in arriving at this knowledge follows an informal pattern of argument that I call *an argument from error* or *learning from error*.
The overarching structure of the argument is guided by the following thesis:

> It is learned that an error is absent when (and only to the extent that) a
> procedure of inquiry (which may include several tests) with a high probability
> of detecting the error if and only if it is present, nevertheless fails to do so.[4]

Its failing to detect the error means it produces a result (or set of results)
that is in accordance with the absence of the error. Such a procedure of
inquiry may be called a reliable (or highly severe) error probe. According
to the above thesis, we can argue that an error is absent if it fails to be
detected by a highly reliable error probe. A corresponding assertion as to
the error's absence may be said to have passed a severe test.

By 'such a passing result' I mean one that accords at least as well with H as e does.
Its complement, in other words, would be a result that either fails H or one that still
passes H but accords less well with H than e does. Often, it is useful to express (SC)
in terms of the improbability of the passing result. That is:

  (1b) *Severity Criterion (SC)*: There is a very low probability that test procedure T would yield such a passing result, if H is false.

[4]In terms of a hypothesis H, the argument from error may be construed as follows:
Evidence in accordance with hypothesis H indicates the correctness of H when (and
only to the extent that) the evidence results from a procedure that with high probability
would have produced a result more discordant from H, were H incorrect.

6. An Analogy With Diagnostic Tools

Tools for medical diagnoses (e.g., ultrasound probes) offer other useful analogies to extract these intuitions about severity. If a diagnostic tool had little or no chance of detecting a disease, even if it is present (low severity), then a passing result — a clean bill of health — with that instrument fails to provide grounds for thinking the disease is absent. That is because the tool has a very high probability of issuing in a clean bill of health even when the disease is present. It is a highly unreliable error probe. Alternatively, suppose a diagnostic tool had an overwhelmingly high chance of detecting the disease, if present — suppose it to be a highly severe error probe. A clean bill of health with that kind of tool provides strong grounds for thinking the disease is not present. For if the disease were present, our probe would almost certainly have detected it.

It is important to stress that my notion of severity always attaches to a particular hypothesis passed or a particular inference reached. To ask "How severe is this test?" is not a fully specified question until it is made to ask: How severe would a test procedure be, if it passed such and such hypothesis on the basis of such and such data? A procedure may be highly severe for arriving at one type of hypotheses and not another. To illustrate, consider again a diagnostic tool with an extremely high chance of detecting a disease. Finding no disease (a clean bill of health) may be seen as passing hypothesis H_1: no disease is present. If H_1 passes with so sensitive a probe, then H_1 passes a severe test. However, the probe may be so sensitive as to have a high probability of declaring the presence of the disease, even if no disease exists. Declaring the presence of the disease may be seen as passing hypothesis H_2: the disease is present. If H_2 passes a test with such a highly sensitive probe, then H_2 has not passed a severe test. That is because there is a very low probability of not passing H_2 (not declaring the presence of the disease) even when H_2 is false (and the disease is absent). The severity of the test that hypothesis H_2 passes is very low.

7. The Roles of Statistical Models and Methods

Experimental inquiry is a matter of building up, correcting, and filling out the models needed for substantiating severe tests in a step-by-step manner. Standard statistical ideas and tools enter into this picture of experimental inference in a number of ways, all of which are organized around the three chief models of inquiry. Three main roles are:

(i) providing techniques of data generation and modeling along with tests for checking if assumptions of data models are met;

(ii) providing tests and estimation methods which allow control of error probabilities, and

(iii) providing canonical models of low-level questions with associated tests and data modeling techniques.

In each case, I readily admit that the functions served by the statistical tools do not fall out directly from the mathematical framework found in statistical texts. Because of this, there are important gaps that need to be filled in by the methodologist and philosopher of experiment.

The three tasks just listed relate to the models of data, of experiment, and of primary hypotheses and questions respectively. (i) The first task involves pre-trial planning to generate data likely to justify assumptions of the analysis of interest, and after-trial checking to test if the assumptions are satisfactorily met.

(ii) The second task centers on what is typically regarded as statistical inference proper; namely, specifying and carrying out statistical tests (and the associated estimation procedures) or informal analogs to these tests.

It is important to emphasize, however, that the error statistical program brings with it reinterpretations of the standard as well as extensions of their logic into informal arguments from error. The criteria for selecting tests depart from those found in classic behavioristic models of testing. One seeks, not the "best" test according to the low error probability criteria alone, but rather sufficiently informative tests.

Accordingly, what directs the choice of a test statistic, and its associated reference set and error probabilities, is the goal of ensuring that something relevant is likely to be learned. Tests are not used as automatic accept or reject rules — accepting or rejecting hypotheses according to whether outcomes fall in the rejection region of a test. Rather, one infers those hypotheses that pass severe tests in the manner just described. The interpretation of results after-the-trial are reinterpreted as well. For one thing, because the severity calculation must be sensitive to the actual outcome reached, it is not enough to know whether the result fell in the rejection region. Secondly, the assertion warranted is rarely one of the preset statistical hypotheses themselves, but more commonly claims about the discrepancies from those hypotheses that are or are not indicated by the data. My position is that the value of the standard tests of preset hypotheses (e.g., a null hypothesis asserting a 0 difference in means) is that they provide the basis for more custom-tailored inferences (e.g., learning the extent to which an effect exceeds 0). A careful look at applications of standard tests in science, I believe, bears this out.

(iii) Experimental inquiries are broken down into piece-meal questions such that they can be reliably probed by statistical tests or analogs to those tests. The questions, I propose, may be seen to refer to standard types of errors. Strategies for investigating these errors often run to type. I delineate four such standard or canonical types of errors:

(a) mistaking chance effects or spurious correlations for genuine correlations or regularities

(b) mistakes about a quantity or value of a parameter

(c) mistakes about a causal factor

(d) mistakes about the assumptions of experimental data.

Statistical models are relevant because they model patterns of irregularity that are useful for studying these errors.

8. Injecting Statistical Considerations

Reliable inferences are made by learning to ask questions by tapping into one of the known patterns of variability. Often this is accomplished by introducing or injecting statistical considerations into inquiries. Statistical considerations are introduced in two main ways: (i) by means of the collection of data, and (ii) by means of the modeling of the data (manipulations on paper). I discuss these in turn:

(i) Suppose one is interested in a quantity μ, say a mean value of a quantity in a population of interest. One way to introduce statistical considerations is to take a random sample of n members from the population and average up their values for this quantity. This would be to observe the value of the statistic — the sample mean. The single value of the mean that would be observed may be viewed as a random sample from a *hypothetical* population consisting of all of the possible n-fold samples of grains that could have been taken, and the mean value of each sample recorded. Why should we be interested in this hypothetical population of means? Because the mean of the hypothetical population (of means) equals the mean of the real population of interest. By collecting data in a certain way (e.g., so that random sampling is approximately satisfied), one gets the sample mean to be related statistically to the population mean in the sense that the mean of is itself equal to the population mean μ. In addition, the variability in the hypothetical population is related in a known way to the variability in the real population. Thus, the observed sample mean can be used to give an interval estimate of μ and one can attach error probabilities to this estimate. Through this trick, learning about the hypothetical population may used to learn about the real population.

(ii) The second type of deliberate introduction of statistics is by way of data modeling or by statistical manipulations "on paper" or on computer. In order to turn raw data into data that can be used to answer questions posed in the experimental observations must be condensed and organized. The idea is to do something that will enable the actual outcomes to be seen as a single random sample from the population of possible experimental outcomes (the sample space of the experimental model). If we can perform this feat, then we can ask of this *single* sample whether it may be seen as a random sample from a population with a given hypothesized distribution.

Needed is a characteristic of the data — a statistic — such that this statistic, whose value we can observe, will teach us about the parent population. The experimental strategy is essentially this: One begins with a handful of standard or canonical models — such as those offered by statistical distributions.[5] One then thinks of ways to massage and rearrange the data until arriving at a statistic, which is a function of the data and the hypotheses of interest, and which has one of the known distributions. Nothing in front of you need actually have the distribution you arrive at. It may simply be the distribution followed by the random variable arrived at through manipulations on paper (e.g., averaging, dividing by or adding appropriate numbers, squaring). But that is all that is needed to assign probabilities to various outcomes on the hypotheses being tested. With this, statistical tests can be run, and their error probabilities calculated. And these error probabilities (e.g., severity) *do* refer to the actual experimental test.

9. Learning What it Would Be Like

The key role for probability distributions of experimental (random) variables, in this account, is to inform us about what *would be expected* under various assumptions about aspects of the underlying experimental process. Whether it is by pointing to a statistical calculation, a pictorial display, or a computer simulation, the 'what would it be like' question is answered by means of an *experimental (or sampling) distribution*: a statement of the relative frequency with which certain results would occur in an actual or hypothetical sequence of experiments.

Such answers are informative because of their links to the statistical tools of analysis (tests and estimation methods) that make use of them. A central use of this information is to what it would be like if various different hypotheses about the underlying experimental process *mis*described a

---

[5] Already in your "tool kit" are a bunch of random variables that have these distributions.

specific experimental process. It teaches us what it would be like were it a mistake to suppose a given effect were non-systematic or due to chance, what it would be like were it a mistake to attribute the effect to a given factor, what it would be like were it a mistake to hold that a given quantity or parameter had a certain value, and what it would be like were it a mistake to suppose experimental assumptions are satisfactorily met. Statistical tests can then be designed so as to magnify the differences between what it would be like under various hypotheses.

After learning enough about certain types of mistakes and the ways to make them show up, it can be argued that finding no indication of error despite the battery of deliberate probing is excellent grounds for taking the error to be absent. To suppose otherwise is itself to adopt a highly unreliable method. It is tantamount to supposing that several, well-understood methods, have deliberately conspired to thwart detection.

In sum, standard statistical models afford very effective tools for approximating the experimental distributions needed to convey "what it would be like" under varying hypotheses about the process generating the experimental data, and the error probabilistic properties of these tools enable this information to substantiate arguments from error. One need not look any deeper to justify their use. Adherence to misconceptions as to what a theory of statistics would have to do to provide a philosophically adequate account of inference, supposing, in particular, that it would need to provide a quantitative measure of the relationship between evidence and hypotheses, has let this interesting and powerful role of statistical ideas go unappreciated.

10. Concluding Remarks

The error statistical account licenses claims about hypotheses that are and are not indicated by tests without assigning quantitative measures of support or probability to those hypotheses. To those E-R theorists who insist that every uncertain inference have a quantity attached, our position is that this insistence is seriously at odds with the kind of inferences made every day, in science and in our daily lives. There is no assignment of probabilities to the claims themselves when we say such things as: the evidence is a good (or a poor) indication that light passing near the sun is deflected, that treatment X prolongs the lives of AIDS patients, that certain dinosaurs were warm blooded, that my 4-year old can read, that metabolism slows down when one ingests less calories, or any of the other claims that we daily substantiate from evidence. What there is, instead, are arguments that a set of errors have been well ruled out by appropriately severe tests.

REFERENCES

Hacking, I. (1992). Statistical Language, Statistical Truth, and Statistical Reason:
    The Self-Authentification of a Style of Scientific Reasoning. In E. McMullin
    (Ed.), *The Social Dimensions of Science.* Notre Dame Press. pp. 130–157.
Mayo, D. (1985). Behavioristic, Evidentialist, and Learning Models of Statistical
    Testing. *Philosophy of Science* **52**, 493–516.
Mayo, D. (1996). *Error and the Growth of Experimental Knowledge.* Chicago: The
    University of Chicago Press.
Neyman, J. (1971). Foundations of Behavioristic Statistics. In V. Godambe and
    D. Sprott (Eds.), *Foundations of Statistical Inference.* Toronto: Holt, Rinehart
    and Winston. pp. 1–13.
Pearson, E. S. (1966). *The Selected Papers of E. S. Pearson.* Berkeley: University
    of California Press.
Suppes, P. (1969). Models of Data. In *Studies in the Methodology and Foundations
    of Science.* Dordecht: D. Reidel. pp. 24–25.

Deborah G. Mayo
Department of Philosophy
Virginia Polytechnic Institute
Blacksburg, Virginia 24061
e-mail: mayod@vt.edu

Commentary by Susan Vineberg

In her paper, "Science, Error Statistics, and Arguing From Error," Deborah Mayo sketches an account of scientific inference that she thinks has several important advantages over the widely held Bayesian view. On the Bayesian view of scientific inference, which Mayo classifies as being an evidential relationship (E-R) approach, confirmation is to be understood in terms of a probabilistic relationship between theory and evidence. For the Bayesian, evidence e confirms a theory T if and only if $p(T|e) > p(T)$. This characterization of the confirmation relation, when paired with the Bayesian Rule of Conditionalization, yields a theory of inductive scientific inference, which is entirely general in that it applies where the relationship between theory and evidence is either statistical or deterministic. However, where the relationship is statistical, it is Bayesian statistical methods that are to be invoked in characterizing when evidence e provides support for a hypothesis.

In contrast, the error statistical approach to scientific inference does not involve assigning probabilities to hypotheses at all. Instead, classical statistical procedures are used to determine the empirical support for a hypothesis. On this approach probabilities refer only to properties of statistical tests or estimation procedures (i.e. error probabilities), where such probabilities are based on objective frequencies. As Mayo characterizes this approach, a hypothesis H is said to pass a severe test with e if (1) e fits H and the statistical test procedure had a high probability, in the sense above, of producing a result that fits less well with H than e, if H were false. Evidence e is then said to confirm H when H passes a severe test with e.

The account of scientific inference that Mayo develops from Neyman and Pearson statistical methods provides an alternative to the Bayesian account, which deserves much more attention than I can provide here. Among its virtues, Mayo claims that the error statistical approach yields an account of scientific inference, which is "truly" ampliative, objective and fits actual scientific practice. Her implication is that E-R approaches, and Bayesianism in particular, are unsatisfactory on all three grounds[1] What I will be arguing in the body of this commentary is that none of these alleged virtues provides a clear and substantial reason to prefer Mayo's error statistical account of scientific reason to the Bayesian one.

Suppose that H is taken as confirmed on the basis of evidence e, in accordance with the error-statistical model of inference, where it is assumed that e does not entail H. From the assumption of the data e, the affirmation of H is clearly ampliative, since by assumption H has content over and above e. Why though would one think that a Bayesian inference from e to the high probability of H is any less an ampliative inference? I suppose that the reasoning might run as follows: The Bayesian's inference begins with an assignment of prior probability to H given e. When e is learned the new probability for H is then taken as the prior probability of H given e, in accordance with Bayes' Rule. From Bayes' rule, the posterior probability for H follows deductively from e together with the prior conditional probability for H given e. However, this in no way shows that the inference from the evidence to the probability of the conclusion is a non-ampliative inference. It is just that, in effect, the ampliative inference within the Bayesian scheme is manifested in the assignment of prior probabilities.

It should be observed that the situation is quite analogous on the error statistical account. The fact that a given H hypothesis passes a severe test with e is a matter of deduction and can be anticipated in just the way that the posterior probability for H can on the Bayesian view. This, of course, does not show that the inference from the data e to H is not an ampliative one.

---

[1]While Mayo only implies that Bayesianism fails on these grounds in the present paper, she elaborates these criticisms in Mayo (1996).

On the Bayesian view, the probabilities that figure in scientific reasoning are to be interpreted as personal degrees of confidence. Accordingly, the view is sometimes called personalist or subjective. This has lead to the misunderstanding that the Bayesian account of scientific reasoning is not objective. Scientific reasoning, which is to be taken on the Bayesian view as proceeding from evidence to posterior probabilities is objective ; however, since there are no general Bayesian constraints on prior probabilities, the assumptions to which Bayesian reasoning is applied need not be objective. However, the fact that there are no constraints on rational degrees of belief beyond satisfaction of the probability axioms that all self designated Bayesians agree on, in no way shows that there aren't additional constraints on rational degrees of confidence for scientific claims.[2] However, to the extent that the prior probabilities employed in Bayesian statistical practice have an element of subjectivity, it should be noted that it is far from clear that the error statistical approach avoids this subjectivity either (for discussion, see Howson and Urbach 1993).

Mayo emphasizes that the error statistical account of scientific reasoning fits actual scientific practice, whereas E-R approaches do not. Mayo is certainly correct that scientists do not typically report probability assignments for the hypotheses that they evaluate. It is also true that scientists generally employ classical statistical methods, which do not involve assigning probabilities to hypotheses. Are these good reasons to reject E-R approaches and the Bayesianism in particular in favor of the error statistical analysis? I claim that they are not. First, while scientists typically employ classical statistical methods, this is surely not the result of a comparative evaluation of Classical and Bayesian methods, but is rather reflects the fact that it is the classical methods that they have been taught. There are Bayesian correlates of many classical statistical methods, which means that in many cases scientists could use Bayesian statistical methods instead (Lindley 1972). It should also be pointed out that some scientists do use Bayesian statistical methods. Indeed, there are cases where classical methods are not applicable, but Bayesian methods are, such as where scientists need to compare the probabilities of two different hypotheses. In particular, Bayesian statistical methods have been fruitfully employed to this end in economics (for details, see Leamer 1984).

Although some scientists consciously employ Bayesian methods, the majority do not. The fact that such scientists do not regard themselves as assigning probabilities to hypotheses does not mean that their attitudes and actions do not show that they are best interpreted as making

---

[2]For instance, Maher (1996) has argued that satisfaction of the probability axioms is not a sufficient condition on rational degrees of confidence.

such assignments. What is crucial is that probability assignments can be attributed to a person, provided that her preferences satisfy certain conditions. It is reasonable to think that scientists do have preferences, which for the most part, allow us to interpret them as attaching probabilities to hypotheses, even though they do not do so consciously (for discussion, see Maher 1993). Herein lies the great advantage of the Bayesian theory. It not only provides an account of scientific reasoning in both deterministic and statistical contexts, but also incorporates such reasoning into an overall theory of rational decision and action.

REFERENCES

Howson, C. and P. Urbach (1993). *Scientific Reasoning: The Bayesian Approach.* La Salle, Illinois: Open Court.

Leamer, E. (1984). *Sources of International Comparative Advantage: Theory and Evidence.* Cambridge, MA: MIT Press.

Lindley, D. V. (1972). *Bayesian Statistics, A Review.* Philadelphia: Society for Industrial and Applied Mathematics.

Maher, P. (1993). *Betting on Theories.* Cambridge: Cambridge University Press.

Mayo, D. (1996). *Error and the Growth of Experimental Knowledge.* Chicago: University of Chicago Press.

Susan Vineberg
Department of Philosophy
Wayne State University
Detroit, Michigan 48202
e-mail: susan.vineberg@wayne.edu

Mayo's Reply to Vineberg

While scientists typically employ classical statistical methods, Susan Vineberg tells us "this is surely not the result of a comparative evaluation of Classical and Bayesian methods, but rather reflects the fact that it is the classical methods that they have been taught" (p. 108). This disparagement of scientists' knowledge and critical abilities, however comforting to subjective Bayesian philosophers frustrated with what they perceive as the stubbornness of scientists, is quite unwarranted. Bayesian methods are taught along side "classical" ones, and both are used when appropriate. What the bulk of scientists who use statistics reject — and quite consciously — is the suspension of objectivity licensed by the brand of subjective Bayesianism embraced by some philosophers. But some subjective Bayesians think it is the scientist who has got it wrong. The fact that prior probabilities solely measure personal degrees of belief, says Vineberg, "has lead to the misunderstanding that the Bayesian account of scientific reasoning is not objective" (p. 108); after all, the reasoning "from evidence to posterior probabilities is objective" (p. 108), being a matter of deductive logic. Such talk (admittedly encouraged by Howson and Urbach) reveals just how divorced some subjectivists have become from the scientific striving for objective, responsible control over assertions. In my brief comments I will focus on the real and serious difference between the error statistical and subjective Bayesian accounts on the matter of objectivity. Mimicking Vineberg, we should ask: Why are so many philosophers subjective Bayesians? In this case, it really is not the result of a fair-minded comparison between Classical and Bayesian approaches but rather that Bayesian philosophers of science have assured them that all "classical," or, as I prefer, error statistical methods have been discredited — despite their widespread use in science. The most unfortunate upshot of this unfamiliarity with the error statistician's tool-kit is that it has prevented philosophers from appreciating how these tools have taken the highroad of objectivity in science. When it comes to satisfying the most fundamental requirements of objectivity, the difference between the subjective Bayesians and error statisticians could not be more stark. Contrast the very aims and goals each assigns to a theory of statistics. While the subjective Bayesian seeks tools for representing and updating an agent's subjective preferences and beliefs, the error statistician seeks tools for checking and uncovering the errors to which personal beliefs and desires can lead. Where L. J. Savage makes it very clear that the theory of personal probability "is a code of consistency for the person applying it, *not a system of predictions about the world around him*" (Savage 1972, p. 59, emphasis added), error statistical theory is a conglomeration of tools for learning about the world around us in a reliable and intersubjective manner. Posterior probabilities may teach about the strength of an agent's beliefs, but because error probabilities are properties of the test procedure as applied to the phenomenon of interest, they can be used as keys to open up answers to questions framed about the phenomenon. While subjective Bayesians use statistical tools to model their beliefs and preferences, error statisticians appeal to them as protection from the many ways they know they can be misled by data as well as by their own beliefs and desires. The differences are radical. If a subjective Bayesian agent has a strong belief in a hypothesis and believes there is no other plausible hypothesis to account for the data, then he is warranted in according it a high posterior degree of belief. The error statistician, by contrast, would not be warranted in inferring a claim unless and only to the extent that it has been put through a reliable experimental test. The subjective

Bayesian's final report is a posterior degree of belief assignment — it is no part of their analysis to control the frequency of erroneous beliefs. By contrast, the error statistician must report enough information to assess (at least approximately) the overall reliability with which the data has been generated and the inference reached. But doesn't the reliability assessment depend upon the particular test chosen? Yes, but it is a mistake to suppose this introduces an obstacle to objectivity: the latitude that exists in the choice of test does not prevent the determination of what a given result does and does not say. The error probabilistic properties of a test procedure — however that test was chosen — allows for an objective interpretation of the results. Even if given test specifications reflect the beliefs, biases, or hopes of the researcher, those factors are quite irrelevant to scrutinizing what the data say and do not say. They pose no obstacle to my scrutinizing any claims you might make based on the tests, nor to my criticizing your test as biased, flawed, or otherwise failing to warrant your inference. Nothing analogous can be said for criticizing subjective degrees of belief. Objectivity in science is directly related to admitting one can be in error, and to agreeing to subject one's claims to a severe scrutiny by others. This calls for intersubjective tools for assessing the evidence and for adjudicating disagreements about hypothesis appraisal on the basis of the kinds of information scientists actually tend to have. This is what error statistical tools provide. Subjective Bayesianism, by contrast, seems to mean never having to say you're wrong (Dennis 1996).

REFERENCES

Dennis, B. (1996). Should Ecologists Become Bayesians? *Ecological Applications* **6**(4), 1100.

Savage, L. J. (1972). *The Foundations of Statistics*. New York: Dover.

*Poznań Studies in the Philosophy
of the Sciences and the Humanities*
2000, *vol.* 71, *pp.* 112–135

Mark Lance

THE BEST IS THE ENEMY OF THE GOOD:
BAYESIAN EPISTEMOLOGY AS A CASE STUDY
IN UNHELPFUL IDEALIZATION

ABSTRACT. Idealization functions in many ways in philosophy. To mention just two, theories can offer idealizations of a subject matter in the sense of producing a simplified model of it which nonetheless bears systematic relations to the actual object of study, or the theory can make various uses of an idealized — i.e. more perfect — subject.

Without denying the values of idealization, this essay attempts to illustrate some dangers inherent in it by way of an examination of one example: the idealizations of Bayesian epistemology. The point is not merely to note that Bayesianism postulates an agent who does not accurately replicate actual epistemic agents. Such is the nature of idealization. Nor is the point merely that this idealization results in a theory which has little to say about important issues in epistemology. It does this, but that point has been well noted by Bayesians. Rather, the claim is that if one makes the idealizations which are crucial to the real explanatory successes of Bayesian epistemology, one is thereby precluded from even raising issues which are arguably at the center of epistemic concern. These issues have largely to do with rational diachronic revision of epistemic states. I conclude with a brief discussion of the general morals which can be drawn from this case study.

"An easy consistency is the hobgoblin of little minds." — Wilfrid Sellars

In the latter part of the twentieth century, a broad range of epistemologists have come to recognize the importance of Wilfrid Sellars's observation, almost four decades ago, that rationality is essentially diachronic. The most famous conclusion of "Empiricism and the Philosophy of Mind" is that both sides of the foundationalist debate lead to an epistemology which is:

> misleading because of its static character. One seems forced to choose between the picture of an elephant which rests on a tortoise . . . and the picture of a great Hegelian serpent of knowledge with its tail in its mouth . . . Neither will do. For empirical knowledge, like its sophisticated extension, science is rational, not because it has a foundation but because it is a self-correcting enterprise which can put any claim in jeopardy, though not all at once (1963, p. 170).

Such an approach has been fruitful. Epistemological reflection on the arbitrariness of any agent's starting point for example reflection on the essential role of trust in our teachers, trust which must in most cases be based on no evidence whatsoever has led to an appreciation that the difference between rational and irrational thought must lie in what one does with these starting points, with how one revises belief in the light of evidence and argument, rather than in the nature of the starting point itself. A related tradition in the philosophy of science, developing the best parts of the work of Kuhn and Lakatos, recognizes that competent scientific practice often considers us warranted in working with theories which are empirically inadequate, lacking in scope, or even inconsistent, so long as the research tradition within which these theories exist is "flourishing," that is, so long as one is able to make sequences of rational emendations of the theory in a non-*ad-hoc* manner.

My central purpose here is not to defend the diachronic perspective in epistemology. Rather I take for granted that it is of at least some importance to be able to answer questions concerning the rationality of belief revision. Given this, I criticize one program in epistemology — Bayesian epistemology — on the grounds that it prevents us from giving such answers. I explain the basic ideas of Bayesian epistemology in §1, emphasizing the idealizations it makes concerning cognitive agents. In §2 I argue that these idealizations are unfruitful. The theory of rationality which must result from looking at this sort of idealized agent is too strong in that it implies normative evaluations of agents which are simply incorrect. The theory is also, and this is the main point, too weak in a particularly significant sense. Not only does Bayesian epistemology offers us only a narrow range of normative constraints on agents, constraints compatible with obviously irrational attitudes, but the very idealizations that motivate what constraints Bayesian epistemology does impose prelude our considering the crucial questions of diachronic rationality in any productive way. It is, I think, fairly clear what bushes we should be beating about in if we wish to search for conditions of diachronic rationality. The problem is that these bushes are just the ones cut down to form the desert landscape that is the Bayesian agent.

Finally, in §3, I look to the question of whether there are general morals to be drawn about the use of idealization in normative theory. It is clear that the lessons of the failures of Bayesian epistemology extend to many other possible projects. It is also clear that these failures do not preclude these idealizations having useful roles in other epistemological explanations. That an idealization prevents one from answering one question is not to say that it should not be allowed to have a role in the answering of any. What is more interesting is the question of whether there are any general conclusions regarding the nature of idealization to be drawn from this case study. I gesture vaguely at some such conclusions.

1. Bayesian Epistemology

Bayesian epistemology begins with the psychological characterization of a deliberating agent which arises in Bayesian decision theory. Such a description begins with the insight that an attitude toward a proposition need not fall into one of the two classes — belief that P is true, belief that P is false — of traditional agentive psychology. Rather, our attitudes are more accurately captured as a numerical range. I might be said to believe that the product of two negatives is positive, that the sun will rise tomorrow, and that Washington will make the playoffs this year. But there is more to it than that. I am virtually certain that the product of negatives is positive, and not much more inclined to doubt that the sun will rise. In the case of views about future NBA results, however, I am really only a bit more confident than not.

Such levels of certainty — degrees of belief, subjective probabilities — are exhibited in my dispositions to accept bets. I would bet just about any amount at just about any odds that the sun will rise tomorrow, but I would not give you 10 to 1 odds on Washington's success. Thus, it seems to the Bayesians that it is natural to represent the psychological state which provides the input to a theory of rational decision to be not a function from propositions to truth values, but a subjective probability function: a function from propositions to points in the closed interval $[0, 1]$.

Now this description is an idealization in several ways. First, it is certainly descriptively inaccurate to say that real people have precise numerical probabilities attached to propositions. I may be inclined to accept bets on Washington giving odds of $3-1$, but not inclined to accept them at $10-1$, thus indicating that my assignment is between .75 and .9, but there may just be no fact of the matter as to whether my subjective confidence is at .8 or .800000001. Similarly, it may be an idealization to assume, as orthodox Bayesians do, that one has any attitude at all toward some propositions. Surely there are certainty claims of the language which I have never entertained at all.

Even more significant idealizations occur, when Bayesianism assumes that one's probability assignment is "coherent," that is, that it respects the axioms of probability theory. One such axiom is a generalization of logical consistency. Just as we might assume of an agent that if she believes P, and if P entails Q, then she believes Q, so Bayesian coherence assumes that if P entails Q, then $p(Q) \geq p(P)$. Of course real agents aren't like this, but the Bayesian idea is to engage in normative idealization. Surely, they argue, it is better to be consistent than not, so that which would rationally be done by a consistent version of oneself is better to do than that which one would do on the basis of inconsistency. So we can, it is thought, learn something important about rational behavior by making the idealization.

Bayesian epistemology begins with this coherence requirement of Bayesian decision theory, a requirement necessary for the formal work of the decision theory to be carried through (though I later consider attempts to weaken various of the idealizing assumptions while staying within the framework.) Coherence already provides one significant necessary condition on rationality. One is epistemically suboptimal, according to the theory, if one's probability assignment violates the axioms of probability theory. Another significant constraint involves the Bayesian account of rational belief change. Given a coherent probability assignment, and given a new bit of information, we can easily see, using a Venn diagram argument of the sort that motivates the axioms of probability, that the new probability for any proposition should be set as follows: $p'(A) = p(A|e) = p(A \,\&\, e)/p(e)$. (Imagine the circles "drawn to scale," that is, showing by the proportion of total space they take up, the likelihood of the proposition they represent. Then the ratio of the size of the $A \,\&\, E$ section to the e section should be $p(A|e)$.)

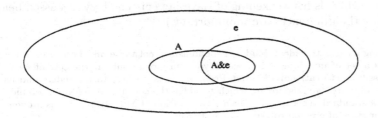

Figure 1.

These two constraints on ideal rationality, synchronic coherence and diachronic change via conditionalization, are made possible by the Bayesian idealizations and allow for some easy solutions to otherwise difficult problems in epistemology and decision theory, such as the lottery paradox, and the preface paradox. Despite this, we now turn to problems.

2. Problems with Bayesian Idealization
2.1. Why Bayesian Epistemology is too Strong

As we saw, Bayesian epistemology requires that a rational agent be probabilistically coherent and if the goal of the theory were *merely* to offer a partial characterization of an epistemically ideal agent, then this wouldn't be problematic. (Though one must wonder why, if this were the only goal, one wouldn't just go all the way and say that an ideal agent assigns probability 1 to every truth and 0 to every falsehood.) The problem comes in when we note the obvious fact that any epistemology is supposed to tell us not just about the best, but something also about the good. We would like an epistemology to give some guidance as to how actual agents ought to reason. Given this, it is often argued that the requirement of coherence is too strict, since it is no epistemic fault to violate coherence so long as the violation is one not something one could reasonably be expected to notice. Thus, considering the probabilistic rule above, that one's assignments respect logical entailment, note that there are any number of entailments which are hard to discover. Given this, it is hardly irrational to assign higher probability, say, to the axioms of set theory than to the four color map theorem in its set-theoretic translation, but since the four color theorem is provable, to do so is probabilistically incoherent. Thus, we see the first suggestion that the best is the enemy of the good.

The response to this charge is fairly standard now among Bayesians, though there are some differences in emphasis. I quote the version given by Mark Kaplan, since it captures the essentials of other similar responses,[1] and has certain advantages. (Note that Kaplan refers to "modest probabilism," which is his weakening of Bayesian epistemology discussed below. For now the differences are unimportant.)

> Being a regulative ideal, Modest Probabilism is not to be understood as giving you a set of marching orders which you can violate only at the cost of being classified as irrational. Modest Probabilism is, rather, to be understood as putting forth a standard by which to judge the cogency of a state of opinion — a standard according to which any violation of Modest Probabilism opens your state of opinion to legitimate criticism.
>
> To say that your state of opinion is open to legitimate criticism if it violates Modest Probabilism is not to say that *you* are open to legitimate criticism in such an event. You can hardly be held open to criticism for violations of Modest Probabilism . . . that are due only to your limited cognitive capacities, limited logical acumen, or limited time (1996, pp. 37–38).

So far, Kaplan's defense could seem to render his position epistemologically uninteresting. If he wants to say that people are not to be held

---

[1] Similar defenses have been given by Jeffrey, Levi, Sobel, and Armendt. The latter is the closest to the details of Kaplan's defense.

accountable in any way for violating the strictures of Bayesian epistemology — that only abstract objects such as states of opinion are — then the view could seem to be nothing more than that described earlier, namely, that an ideal set of beliefs would be coherent. (Again, why not be bold: an ideal set of beliefs consists of truths.) This would make the view correct at the cost of eliminating any connection to the normative evaluation of actual agents and their cognitive states. This is not, however Kaplan's intent:

> All the regulative ideal demands of you is that you acknowledge that, to the extent that a state of opinion *does* violate Modest Probabilism, it is open to legitimate criticism.
> An empty demand? Hardly. The question of what constitutes a legitimate criticism of an opinion is central to the enterprise of rational inquiry an enterprise whose aim, after all, is to determine what opinions may be sufficiently free of legitimate criticism to warrant adoption. Modest probabilism offers a small, but significant, part of the answer to that central question (1996, p. 38).

Thus, the point seems to be that whereas retrospective criticism of an agent as irrational in virtue of having violated coherence unknowingly may not be appropriate, what is appropriate is that we require rational agents to take incoherence to be a legitimate criticism of their cognitive states. If someone shows you that your state is incoherent, she has thereby placed an onus of change upon you. You must now revise your degrees of belief or be fairly judged irrational. This is, however, still too strong.

Consider as a representative example a chess master considering a position and attempting to decide if white has a win, that is, if there is a winning strategy for white. This is a standard question which arises in the evaluation of chess games,[2] and it is one that chess masters are quite good at answering. They are good at it because they have good judgment and employ powerful theories having to do with such things as piece activity, pawn structure, material advantage, space, etc. So let us define as follows:

Let sentence (1) be a complete description of the position under consideration.

Let sentence (2) be a conjunction of the rules of chess.

Let (3) be 'White has a winning strategy'.

---

[2] Note that the issue is certainly not typically to be described as having to do with predictions about which player will win. Sometimes chess masters may consider such issues, thereby bringing in questions of the sorts of mistakes likely to be made by each player, the time pressure that each is in, how tired one or the other is, etc. But the common issue is the objective one of what the position dictates, of which side has a winning strategy.

Now let us say that a group of masters have examined the position. They assign probabilities very near to 1 to (1) and (2), of course, and upon examination, assign a probability of .8 to (3). That is, they come to be fairly sure that white is winning, but due to the complexity of the situation, they are not entirely sure, certainly not as sure as they are of the position or the rules of chess.

Now note the crucial fact. The conjunction of (1) and (2) either entails (3) or entails the negative of (3). That is, if white has a winning strategy, then this fact is entailed by the rules and the position, and if not, then this fact is. We could suppose, to make the matter dramatic, that the masters are wrong. Maybe through some complex, counter-intuitive combination which requires going against conventional theory and judgment black is actually winning. But this is inessential, for the above entailment implies that probabilistic coherence either requires one to assign a probability near 0, or assign a probability near 1. If one were to be certain of the rules of chess and the position, the requirement would be to assign 0 if white actually lacks a winning strategy, or to assign 1 if there is one. For simplicity, I will suppose this is the situation in what follows. Since we know that there either is or isn't a winning strategy — even a constructivist knows this, since chess is a finite, albeit really big, game — we know that the masters' cognitive state is probabilistically incoherent. It is never rational to assign a probability between 0 and 1 to such a claim as (3).

Thus, according to Kaplan's line, the opinion of the masters can be criticized, and presumably thereby is not "sufficiently free of legitimate criticism to warrant adoption" (p. 38). But this is absurd. Their opinion is a better one than that of anyone else. They are, after all, the masters. Nor will it help to move to higher order probabilities, saying that they assign .8 probability to the view that the rational probability of white having a winning strategy is 1. For the conjunction of (1), (2) and Bayesian epistemology entails not just that white has no winning strategy — let us assume this is the case — but also that the probability his having one is 0. So one could assign an intermediate probability to the probability judgment only by assigning low probability to the epistemology. (And in other cases, we will push the requisite confidence in the epistemology down to .5, or worse, to various different values.)

In reality, one would be epistemically *worse* to assign probability 1 to the true claim regarding white's prospects here. One is better to do the only thing that a rational person can do, namely follow the arguments of the experts and conclude that white is quite likely in a position to win. Though there is certainly a possible agent more rational than the masters who would assign the proper extreme value to (3). Again, the omniscient chess genius is an example. But it is obviously false that an expert but fallible agent would become more rational by coming to share this one property of probabilistic consistency with the chess player none better than which can be imagined.

The general point is one which applies to any sort of coherence theory whether Bayesian or not. Any view which requires the rational to avoid inconsistency — even to avoid inconsistency which can be shown to obtain — is, in an important sense, externalist. The reason is that the logical properties of reasonably complex systems — even big finite ones like chess — are no more transparent to the mind than are, for example, the causal pedigrees considered by reliabilists. Thus, all the worries which beset externalists — worries having to do with cases in which the external property obtains, but one has no reason to think it does, or even has positive reason to doubt it — beset consistency advocating coherentists as well.

Before ending this section, it is worth noting that there is a claim in the vicinity of Kaplan's which has at least some plausibility. This is the claim that if one is confronted with an actual violation of a particular axiom — e.g. if one is shown that A entails B, and that she assigns lower probability to B than to A — then one is required to alter the assignment in some way. Such a demonstration would seem to be a legitimate criticism — though even here there is room for worry[3] — but acknowledging this is far from the Bayesian notion. It is far precisely because it is internalist. It is having been shown, coming to know, that one is violating a particular rule that requires one to come into conformity with it. Only when the relevant region of logical space is made transparent are we required to bring our cognitive states into accord with it. Merely being such as to violate a rule, or even knowing that one is violating some rule or other, places no onus on one to come into conformity with the rule one is violating.

2.2. Why Bayesian Epistemology is too Weak

There is a fairly innocuous sense in which, as has often been noted, Bayesian epistemology is too weak. That is, reflection on the constraints

---

[3] Say that I am shown that A entails B and that I am fairly sure of A and doubt B, in a case in which the connection is quite difficult. Maybe I accept the axiom of choice, but I am inclined against the well ordering postulate. There may, then, be no good grounds for deciding whether to lower the probability of A (reject it) or to raise that of B (accept it). Until such is forthcoming, could it not be rational to attempt to go ahead and use the inconsistent pair of claims each in their respective domain, attempting to avoid contamination of one by the other, and awaiting future clarification? Perhaps so, but I think one must at least acknowledge that this sort of demonstration is a failing in one's cognitive state, and a failing which will have significant consequences for how one goes on to reason.

imposed by probability theory simply don't take one too far. After all, there are all sorts of just plain crazy probability assignments which are coherent, (e.g. ones which assign every atomic probability 1). The response to this is easy to anticipate, however. Of course there must be more to say, but Bayesian epistemology never purported to offer more than a set of necessary conditions on rationality. Further, the necessary conditions it imposes are certainly significant. Not only the lottery and preface paradoxes mentioned above receive a resolution, but a range of confusions which actually occur in reasoning can be seen for the errors they are when approached within the Bayesian framework.

There is, however, a deeper sense in which Bayesian epistemology is too weak. Not only are certain questions not dealt with by the theory as developed, but that development actually precludes the possibility of addressing them in promising ways. To see this, we turn to the Bayesian account of diachronic rationality. We begin with an argument of Van Fraassen's against the epistemological principle of inference to the best explanation (IBE). The argument is clearly a bad one, specifically a straw man argument, but seeing this will allow us to recognize that in the case of diachronic principles, conditionalization must be the entire Bayesian story. Thus, if the account is incomplete, it cannot be completed.

Van Fraassen (1989) criticizes the defender of IBE by first representing the view in the Bayesian framework. To think that one should care about whether an explanation is best, according to Van Fraassen (1989), is to think that when confronted with new evidence, one ought to assign the best explanation of that evidence a probability higher than that which one otherwise would assign it, namely a probability higher that the conditional probability of the explanation given the evidence. From this assumption, Van Fraassen constructs a "Dutch Strategy" argument. Roughly, he argues that any person adopting such a strategy ought to be prepared to accept a set of bets which insure that no matter how things go the agent will lose money. Thus, the having of such a strategy is tantamount to a set of inclinations which are logically guaranteed to lose.

The details of the argument don't matter, since they are correct. Before turning to what I take to be the proper response to the argument, I consider a few responses which are not satisfactory. Van Fraassen himself accepts this argument and takes the conclusion to show that one cannot rationally adopt any diachronic strategy other than conditionalization, but urges that we not adopt strategies at all. The Dutch Strategy argument only goes through if one assumes the agent has a strategy of inferring by IBE. Merely arguing in a manner contrary to conditionalization on a given occasion does not logically guarantee a loss according to the set of bets, so according to Van Fraassen one is rationally allowed to infer differently, maybe even always in accord with IBE, so long as one does not adopt a strategy of doing so. Now it would be easy to think ill of an epistemology that so required one to eschew principled reflection on one's diachronic behavior, but we note for now simply that the view does involve claiming that a Bayesian agent can rationally infer other than by conditionalization.

A similar conclusion is reached by Mark Kaplan, who does not accept the Dutch Strategy argument. Again, the details need not concern us, but Kaplan rejects this argument as showing anything important about rationality. He combines this with positive criticism of the requirement that one revise by conditionalization. One particularly damning consequence of conditionalization is that it can never lead to a revision of conditional probabilities themselves. Suppose we start with a probability assignment p, and move to a new assignment p' by conditionalizing on e. Then $p'(A|e) = p'(A \& e)/p'(e)$. But since $p'(e) = 1$, this just gives us $p'(A)$ which is $p(A|e)$. So conditional probabilities don't change by conditionalization, but given that the conditional probability of A on e — the verdict regarding A that one is going to have should one learn e — amounts to the Bayesian measure of how much rational evidential support e provides for A, this result is tantamount to assuming that one's judgments of evidential support are unrevisable by evidence. For this and other reasons, Kaplan too wants to deny that conditionalization is mandatory.

Neither position seems satisfactory, however. We needn't rely on the Dutch Strategy arguments in defending conditionalization within the Bayesian framework, for we have the much simpler, and to my mind utterly compelling, Venn diagram argument given above. This argument just shows that one who has the relevant probability assignments is already committed to a view as to how to revise $p(A)$ when she learns that the world is actually within the e space. Given such a set of antecedent opinions, what could possibly count as reason, even as motivation, for changing in any other way?

Susan Vineberg has recently tried to avoid this conclusion. Vineberg (this volume) argues that we should treat conditionalization as a logical principle, akin to a valid conditional in ordinary logic. The epistemic import, then, of the conditional is disjunctive. When one believes A and learns that A entails B, one is required either to accept B or to revise A, but neither of these is required *simpliciter*. Similarly, a Bayesian agent *could* reasonably confront new evidence by conditionalizing, but we must recall that crazy prior assignments are allowed by Bayesian strictures. So one might be in a situation in which new evidence is better seen as an opportunity to revise one's prior assignment function, to remove some of the craziness.

Again the problem is what could possibly motivate such a revision. Of course a Bayesian agent could simply *change* to a new assignment after learning that e, but what would make such a change rational, specifically, what grounds could there be for a change to one revision rather than another? Recall, that on the Bayesian characterization of the situation, one already has assigned a probability to each sentence of the language. Further, none of these assignments is probabilistically incompatible with any other. By definition, the only revision motivating change is one's coming to learn e. So one has a settled attitude toward each proposition, none conflicts with any other, and these attitudes imply a commitment as to the likelihood of A in that region of possibility space in which e is true. What, in such a situation, could motivate one to do anything but conditionalize?

This is not to say that people shouldn't change crazy assignments of probabilities into non-crazy ones. One should, but the reason is that the assignment is crazy. The learning of new evidence does not give a Bayesian agent the reason to change. New evidence, even if the evidence is of the form 'my priors are crazy', can make the craziness of a set of priors neither more nor less than it was before for a Bayesian agent. If one has a complete probability assignment to the language, and learns 'my priors are crazy in way c', one ought simply to revise by conditionalizing on 'my priors are crazy in way c', a strategy already envisioned in the prior assignment of probabilities. That is to say, that while Vineberg can certainly offer up versions of craziness and advocate non-crazy coherent probability assignments over others — thereby giving additional substantive probabilistic conditions of synchronic coherence — this will have nothing to do with belief change in the face of evidence. It cannot be formulated within the Bayesian framework as a diachronic rule.

A similar moral applies in the case of Van Fraassen's criticism of IBE. It is a caricature to attribute to the advocate of IBE that a person who has a commitment to the relevant probabilities, i.e. someone who is committed to taking the conditional probability of A on e to be n, would revise in any other way. Of course if one assigns $p(A|e) = n$, then she will be inclined to assign $p(A) = n$ upon learning that e. The only reasonable interpretation of IBE in this framework is as an additional constraint on priors. Thus, one ought, according to this Bayesian IBE, assign $A \& e$ a higher value than any other sentence $B \& e$, if A is the best explanation of e together with existing data. IBE, then, would just be one of the additional constraints on coherence which Bayesians are so quick to admit must exist.

The upshot of all this discussion is the conclusion that conditionalization is rather different than the other logical constraints on belief assignment proffered by Bayesians. Where the axioms of probability provide constraints on rational belief assignments, but do not purport to be sufficient conditions on synchronic rationality, conditionalization is the entire story. There can be nothing to diachronic rationality in the face of evidence other than conditionalization for it can never be rational, for a Bayesian agent to revise in light of evidence other than by conditionalization.[4] We in the meta-language may say that an agent ought *to be* different. That is, we might judge a coherent set of opinions to be defective in that they assign the best explanation of a range of data a lower probability conditional on that evidence than an inferior explanation, but we can't turn this into a rule of revision to be used by the agent herself. Assuming only that she has an assignment already covering all the claims of the language — and here that would include claims about her own probability assignments, about what is the best explanation of what, etc. — it could not be rational, upon coming to learn any claim, to do other than conditionalize on it.

Now this might not seem such a problem, for the more charitable IBE which I offered in response to Van Fraassen's argument may seem to point to a way of reading diachronic rationality back into the synchronic structure of a belief assignment. That is, one could think of the moral as being that when a traditional epistemologist proposes a diachronic rule to the effect that one should move from state S to state S' as a result of coming to accept e, the Bayesian should regard this simply as an additional constraint on the prior conditional assignment of a rational agent. This attempt to synchronize diachronic rationality simply won't work, however.

One reason is that one doesn't want to render irrational, or even judge epistemically sub-optimal given the real situation inhabited by finite logical beings, every assignment which fails to code within its priors every change that it would be rational to make given appropriate evidence. Thus, looking back at the chess example, we want to say that one *should* assign (3) probability 0 upon learning that white has no winning strategy, i.e. upon learning that (1) and (2) entail not-(3). But this entailment, being a logical truth, already must be assigned probability 1 by any coherent agent, so conditionalizing on it is trivial, changes no probabilities.

Similarly, we still have Kaplan's problem, that no conditionalization can change a conditional probability, whereas we do want to say that one ought to be led by new evidence to change one's conditionals. That is,

---

[4]Throughout I have been formulating the discussion in terms of simple conditionalization. Nothing changes in the argument if we switch to the more interesting context of Jeffrey conditionalization.

one should be such that learning e' can change one's inclination to change confidence in A in light of learning e. Yet there simply is no coherent way to represent such a diachronic principle in the synchronic structure of a coherent Bayesian agent.

Finally, note that real people often have no attitude whatever toward many propositions. There are propositions one has never even entertained, or that one is not even capable of understanding. (As Kaplan correctly argues, this should not be modeled as if one assigned probability .5 to the proposition.) Noting this, suppose we take the root notion of a conditional probability to be one's commitment regarding the probability of A upon learning that e. Then we can note that it is quite common to have a settled idea as to the conditional probability of A given e without having any idea of the reasonable likelihood of $A \& e$, though on the Bayesian scheme the latter is definitive of the former. Suppose that I am told by the infallible neighborhood oracle — an oracle I trust completely on the basis of the usual sorts of divine inspiration and empirical wonder — that there are two buried urns, one in my yard, and one in my neighbor's. One of the urns, I am told is red, and contains 4 white balls and 6 black. The other is green and contains 6 white and 4 black balls. The oracle also tells me that she will tell me tomorrow what color the urn in my yard is and allow me to pick a ball at random from that urn after digging it up.

In such a situation, it is perfectly reasonable to hold that p('the ball chosen from the urn in my yard is white'—'the oracle tells me that the urn is red') is .4. This, the orthodox Bayesian could rightly infer, implies that *if* I assign a probability to 'the oracle will tell me that the urn is red' and to the conjunction of this sentence and 'the ball chosen from the urn in my yard will be white', then the result of dividing the probability of this conjunction by the probability of the claim that the urn is red ought to be .4. What shouldn't be assumed is that I must have such an assignment to the conjunction, or to the evidence claim. I have no idea in this situation how likely it is that the oracle will say one thing rather than the other about the color of my urn.[5]

Bayesian epistemology offers us two sorts of constraints on rationality, both of which have correlates in ordinary belief/desire epistemology and decision theory. In each case we revise the more traditional notions in ways which follow from the central Bayesian transition from a truth value assignment to a probability assignment. First, where the traditional epistemologist uses logic to characterize consistency among truth value assignments, the Bayesian uses probability theory to define the richer notion

---

[5]Kaplan has something to say about this possibility of lacking an attitude toward a proposition, but it is not an answer which is much help with the sorts of problems we are developing here. I discuss Kaplan's weakening of orthodox Bayesianism below.

of probabilistic coherence. Similarly, there is a correlate of the belief revision principle of modus ponens, according to which if one is committed to $A \to B$ and then comes to learn that A, she ought to revise her assignment to include B. This principle is extended into the context of degree of belief structures as the rule of conditionalization. Important theorems can be proved and explanations given on the basis of each of these conceptual enrichments.

The problem is that most of the interesting work in diachronic epistemology involves belief revision which does not proceed by this mechanical Bayesian modus ponens. Though I offer nothing like a theory of rational belief revision here, I suggest some broad features of rationality in the case of interesting revisions, i.e. belief revision complicated enough not to be mechanical in this obvious way. Such revision falls into two sorts. One the one hand, it often proceeds by ferreting out inconsistencies in our existing attitudes, inconsistencies which we are made aware of, either by way of logical argument, or by running into an empirical datum which contradicts an *a priori* conclusion.[6]

The problem is that the nice theorems of synchronic Bayesianism require that we idealize away from just those inconsistencies. It is clear that we do make intellectual progress by finding and repairing incoherencies in our cognitive system, that the difference between rational and irrational individuals often consists in how well they do this over time, and that it is the job of epistemology to illuminate how this is properly done. However, all the theorems of synchronic Bayesian epistemology require idealizations which prevent our even raising these questions. If synchronic Bayesianism is *part* of the story, no answer to the question of how best to deal with inconsistency can be another part.

The other class of revisionary strategies involves confronting new propositions toward which we have no attitude whatever. In light of new evidence which calls our prior attitudes into question, one often must confront propositions never before entertained, thus assigning probabilities to new sentences and inferring changes throughout the cognitive fabric on the basis of these enrichments. There are several species of this linguistic supplementation strategy. Sometimes we require new vocabulary: new logical connectives, such as relevant conditionals, new mathematical devices such complex geometric coordinates, or new theoretical posits. On other occasions, we supplement the language by making conceptual distinctions, e.g. between rest mass and relativistic mass. Sometimes we redraw the conceptual boundaries in more complex ways, for example re-

---

[6]I have in mind here one who is convinced by Zeno arguments, yet notices that things move. More sophisticated cases are easy to find in the history of science.

placing the distinction between water-based and land-based animals with that between fish, mammals and others.[7] All these cases involve revisions which begin by considering new propositions to which no attitudes are attached antecedently, and then continue — here being the crucial point — by inferentially connecting the results of this consideration to existing attitudes. Thus, existing beliefs regarding chemical dispositions change upon positing the existence of atoms.

Again, Bayesian epistemology precludes our even considering these issues. If conditionalization is to be part of the story, we must start with a probability assignment to all sentences of the language, since otherwise it is undefined. Much of the action, however, takes place in that region of the language for which there are no priors. Conditionalization cannot even be part of the story, for if it is, then we must idealize in ways which make it impossible to add anything more.

Before turning to some general morals regarding idealization, let us briefly consider Kaplan's "Modest Probabilism," for this is a version of Bayesian epistemology which allows us to weaken some of the idealizations I have identified as problematic for a theory of rational belief revision. Modest Probabilism identifies the state of an agent not with a single coherent probability assignment, but with a set of them. This allows Kaplan to weaken two crucial idealizations: namely to allow that we not have a precise value assigned to a given proposition and that we have no attitude whatever to some. The idea is that what we are committed to is the rational probability of a proposition lying within a set. So if the set includes functions which assign A every value between .4 and .6, we are represented as committed to the probability of A being between these two bounds. If the various functions include all assignments between 0 and 1 to A, then we are represented as having no attitude toward A.

While this certainly makes the theory more realistic, in that it doesn't require one to adopt a precision one has no rational ground for assigning, it does not help with the problems of concern here. The first reason is that it does not at all help with the objection that Bayesian epistemology is too strong. Every function in the set will be coherent, so there is no way to represent incoherence.

In addition, each of these probability assignments must update on the basis of conditionalization, according to the Venn diagram argument. Thus, one's conditional probability for A on e, given a range of prior functions, must just be seen as lying within the range of conditional probabilities for A on e given by the various functions in the range. In the

---

[7]Susan Vineberg pointed out in conversations this species of the general strategy of linguistic refinement.

case of a proposition about which we have no commitments, there will be in the range functions assigning 0 to A and functions assigning it 1. Conditionalizing according to either of these functions will not change the value for A. So these options will stay with us no matter what the conditionalization. And Kaplan's own worry still applies. Conditional probabilities themselves can change in none of the assignments. So the range of acceptable views about evidential strength cannot change.[8]

What we wanted was not just a representation of a non-specific attitude, but one which would allow us to characterize evidential connections between propositions one had no prior attitude towards and propositions one had considered. We want to be able, upon reflection, to consider new claims, assign them values on the basis of our new evidence, and then use them to motivate revisions of our previous verdicts. Kaplan's representation of uncertainty within the Bayesian context still precludes the uncertainty from playing this epistemic role.

3. Morals for Acceptable Idealization

This case study should make us more cautious about the use of idealization in philosophy, specifically in normative theorizing. There is a distressing tendency among philosophers — and certainly it is a tendency not exclusive to Bayesians — to invoke the notion of idealization as a sort of totem against criticism. One offers a way in which a model is inaccurate and is brushed off with the report that *of course* the model is an idealization. Typically this defense is combined with a reference to the ideal gas law, which the objector is supposed to recall from High School, to recognize as an undeniably useful moment in scientific history, and to relate immediately and unproblematically to the present case. I suggest that the preceding case study shows that we must think much harder about such moves.

---

[8]Kaplan himself might want to resist this argument. I argued that a Bayesian agent must revise by conditionalization. One could accept this and still argue that Modest Probabilism represents not Bayesian agents, but agents committed to placing probabilities into ranges. Thus, there is no question of how the various component assignments ought to change. What changes on the basis of evidence is the range to which one assigns a proposition. If this idea could be made to work, we might have some notion of revision within Modest Probabilism which is not constructed out of conditionalization. But that is to say that the view has nothing at all to say about diachronic rationality. By central argument has been to urge that *if* conditionalization is part of the story, nothing else can be added, and so, that we miss out on all that is important in diachronic rationality. Further, this move leaves Kaplan's epistemological position still open to the arguments against the synchronic Bayesian constraints.

We must think hard, in part, because such moves are not simply disingenuous dodges. Indeed, it is quite obvious that one ought in some cases to idealize. Indeed, I would be prepared to argue that precision is impossible without idealization, but it is equally clear that there are many sorts of idealization, and many ways that idealizations can go wrong. Idealization is far from a unitary or a risk-free phenomenon, and it does not in all ways work the same in descriptive and normative theorizing. Though I have nothing like a theory of idealization to offer here, I do want to draw three general morals from the preceding discussion. All are quite elementary, but nonetheless frequently ignored in various areas of the philosophical literature.

First we must keep in mind that an idealization which is helpful in generating explanations of one sort of phenomenon, may not be helpful — indeed, may be positively detrimental — in the search for explanations of otherwise closely related phenomena. Thus, nothing I have said should be taken as casting doubt on the value of the Bayesian explanation of many epistemological puzzles. I have mentioned the solutions to the lottery and preface paradoxes, and could add other successes. Clearly much explicitly probabilistic reasoning is clarified by Bayesian theory. Also, there is a literature which considers our epistemological task to be driven by two competing goals, the avoidance of error, and the achievement of significant and interesting truth. This literature attempts to analyze the project of working for these goals in cost-benefit terms utilizing the resources of Bayesian theory, and has in my view made substantial progress. Still, when we turn to closely related issues, as I have argued, the idealizations which proved so helpful here become liabilities.[9]

The second lesson is more specifically tied to normative idealization. Here it is crucial to keep in mind that normative ideals have a mode of evaluation not analogous to anything in the case of descriptive ideals. A descriptive idealization, such as the ideal gas law, offers laws governing the behavior of an idealization of ordinary gases. This ideal is basically a simplification in one way or another.[10]

---

[9]Perhaps the case is in some ways analogous to the period in science in which we operated with two conceptions of light, as a wave and as a particle. Though this analogy — so far as it goes — suggests that a synthesis would be nice, it would certainly have been wrong to reject either idealization out of hand for all purposes on the grounds that another was necessary for others.

[10]Notice that we simplify in very different ways in different contexts. In the ideal gas law, we attribute simplified properties to each element of the system, supposing that collisions are perfectly elastic for instance. We often do this because we have no idea how to understand the actual nature of the properties, as in the case of real molecular collisions which are both inelastic and most difficult to characterize. In other cases, such as Newtonian derivations of the motion of planets, we ignore many elements of

Now there are, of course, any number of ways to simplify a physical system, and the measure of the quality of one simplification over another — that which makes the ideal gas model ideal and the model of gas molecules as irregularly shaped miniature gummy bears merely simple — is the approximate predictive accuracy of the former model for actual systems. It is that it allows us to obtain approximate results for the expansion of actual gases, together with certain continuity properties of such predictions,[11] that makes the ideal gas model a good one. Corresponding to this virtue of a simplified model is that had by a normative model which allows us to draw approximate conclusions regarding what actual agents ought to do. A model of moral behavior is ideal in this sense if reasoning about the model allows us to draw approximate moral conclusions about how actual people ought to behave. An ideal epistemology, would give approximately accurate advice about how we ought to reason.

In the normative case, however, there is another mode of evaluation which can cut across this one. Normative models can also be evaluated according to the very normative criteria they are supposed to explicate. Thus, typical ideal agents in morality are *morally* better than actual agents. (Though Rawlsian agents, for example, are merely epistemologically better than real people.) Similarly, we could prefer one epistemic idealization on the grounds that it is a model of a more rational agent than is another. What is crucial is not to conflate these two versions of ideality.

Finally, I note that how well a model scores in either of these measures is not something to be determined a priori. New information about a subject matter brings with it new conclusions regarding the additional questions it is most important to be asking. Then, as it has been my goal to argue, changes in the range of questions which are salient can imply changes in the evaluation of the efficacy of various idealizations. When developments in epistemology show that we should attend to belief revision, rather than merely to synchronic relations among beliefs, we learn that many existing idealizations of the epistemic situation — not merely the coherentism and foundationalism criticized by Sellars, but Bayesian epistemology as well — prove counter-productive.

---

the system — e.g. the scattered miniscule particles of space dust — and simplify our description of others — supposing the planets to be perfectly spherical — not because we don't know how to deal with the reality, but because it would be computationally too complex to attend to them.

[11] Ron Laymon (1982, 1987, 1995) has discussed in some detail the importance of a prediction on the basis of idealization not merely being approximately accurate, but also such that small changes in the experimental situation only result in small changes in the level of deviance from the predicted result. More recently, Nancy Cartwright (1995) has discussed cases in which this community fails.

This is hardly a comprehensive list even of the morals which could be drawn from this one case study. Nor are these three morals given in any sort of detail. What I hope is clear is that we ought to turn our collective attention to the sorts of issues of normative idealization they bring up.[12]

REFERENCES

Cartwright, N. (1995). False Idealization: A Philosophical Threat to Scientific Method. *Philosophical Studies* **77**, 339–352.

Laymon, R. (1982). Idealization and Testing of Theories. In P. Achinstein (Ed.), *Observation, Experimentation & Hypothesis in Modern Physical Science*. Cambridge: MIT Press. pp. 147–176.

Laymon, R. (1987). Using Scott Domains to Replicate the Notions of Approximate and Idealized Data. *Philosophy of Science* **54**, 194–221.

Laymon, R. (1995). Experimentation and the Legitimacy of Idealization. *Philosophical Studies* **77**, 353–375.

Kaplan, M. (1996). *Decision Theory as Probability*. Cambridge: Cambridge University Press.

Sellars, W. (1963). *Science, Perception and Reality*. London: Routledge and Kegan Paul.

Van Frassen, B. (1989). *Laws and Symmetries*. Oxford: Clarendon Press.

Mark Norris Lance
Department of Philosophy
Georgetown University
Washington, DC 20057-1076
e-mail: lancem@guvax.acc.georgetown.edu

Commentary by Leszek Nowak

1. Prof. Lance's intriguing case study deserves attention also from the standpoint of the general methodology. The Author draws from his analysis some morals without, however, pretending to explain them in

---

[12]I am grateful to a number of philosophers who have discussed with me issues concerning Bayesian epistemology specifically and idealization in epistemology generally over the years. Ron Laymon introduced me to the importance of considering criteria of reasonable idealization in a course some 15 years ago. More recently, I have discussed issues of Bayesian epistemology and decision theory with Mark Kaplan and Brad Armendt, both of whom taught me a good deal about the subject. I also thank Ken Gemes, John O'Leary-Hawthorne and Joe Camp for various insights on both scores which found their way into this paper in one form or another. Finally, the discussion of the sense in which coherence theories in general can be seen as analogous to externalist conditions on rationality arose in a discussion with students in my graduate epistemology proseminar at Georgetown University.

terms of the general theory of idealization. Since his observations seem to be usually correct and always interesting, I would like to try to explain some of them — ones I take to be true — in terms of some conception of idealization, called the idealizational approach to science (below, briefly, IAS). As to some others, I shall discuss them critically. At the same time, I treat the possibility to explain/to assess these observations as a kind of test for that approach. In this way empirical studies on idealization may meet with general approaches to that procedure. Needless to add, the IAS[1] can not be presented in a concise note and the reader is assumed to be acquainted with its basic notions (idealizing condition, factual and idealizational statements, concretization, influence, etc.) and main hypotheses.

2. Prof. Lance most correctly observes that "[t]here is a distressing tendency among philosophers. . . . to invoke the notion of idealization as a sort of totem against criticism" (p. 127). Take the simplest scheme of a factual statement:

$(t*)$ (x) (if $G(x)$, then $F(x) = f(H(x)))$,

and assume that an object a falsifying $(t*)$ has been found, that is: $G(a)$, $F(a) = \mathbf{n}$, $H(a) = \mathbf{m}$, $\mathbf{n} \neq f(\mathbf{m})$. Now, the simplest way to avoid the objection would be to say: "well, $(t*)$ is nothing but an idealization." Indeed, by adding an arbitrary idealizing condition: $p(x) = 0$ one obtains an idealizational statement:

(t) (x) (if $G(x) \& p(x) = 0$, then $F(x) = f(H(x)))$,

which is vacuously satisfied in the empirical domain G and thus to save (t) against criticism. However, according to the IAS a recourse to an idealizing assumption is admissible on the condition that this very assumption is waived and the appropriate statement is concretized:

(ct) (x) (if $G(x) \& p(x) \neq 0$, then $F(x) = f'(H(x), p(x)))$

to the effect that the initial discrepancy with the experience will be removed (practically, significantly diminished). This holds, if it appears that the following takes place: $G(a), F(a) = \mathbf{n}$, $H(a) = \mathbf{m}$, $p(a) = \mathbf{k}$, $f'(\mathbf{m}, \mathbf{k}) = \mathbf{n}$ (in scientific practice, the approximate identity suffices). It is thus visible that "how well a model scores . . . is not determined a priori" (p. 129). To put it in a rule: you are entitled to defend against an empirical criticism through idealization, on the condition that you successfully attack the same data through concretization.

---

[1]Cf. the monograph (1980). The paper (1992) contains a survey of contributions of various authors.

3. Another correct observation is that ". . . an idealization which is helpful in generating explanations of one sort of phenomena, may not be helpful . . . in the search for explanations of otherwise closely related phenomena" (p. 128). Indeed, according to the IAS, to explain phenomenon (parameter) F is necessary to recall a hierarchy of influences of F in the range G. In case of the idealizational statement (t) it will be the sequence of parameters $\langle H, p \rangle$. However, it may easily happen that the same phenomenon in a "neighboring" range G' undergoes another hierarchy of influences. (Think, for instance, of differences in influences operating on paramagnetic, ferromagnetic and antiferromagnetic materials.) Then F in the range G' may have another hierarchy of determinants, say $\langle R, H \rangle$, and the appropriate idealizational law is, correspondingly, quite different:

(T) (x) (if $G'(x) \& H(x) = 0$, then $F(x) = g(R(x))$).

This problematic is expressed in the IAS as that of relationships between homogeneous (with one essential structure related to the whole universe of a magnitude) and unhomogeneous (with many essential structures related to corresponding ranges of a magnitude) parameters from one side and between the essential structure of a magnitude and its image formed by a researcher from the other.

4. The Author applies to idealization also the standard distinction between description and prescription. My general doubt is about the sharp applicability of that distinction to the realm of idealization. Even a superficial glance suffices to state that ideals (e.g., ideally just society) are abstracts similarly as scientific ideal types (e.g., mass-points, ideal gases, closed economies etc.) are. The only difference between the two lies in human attitudes towards them: the latter are indifferent for us whereas the former are not.

Let us take the simplest definition: a statement "p" is a value-judgment iff there is a group g of people having an attitude of approval (*resp.* disapproval) towards the fact that p and an utterance of "p" is normally taken in g to be a symptom that expresses this attitude. As a result, value-judgments are of two sorts. Some are pure expressions of approval (*resp.* disapproval), e.g. 'That is extraordinary!' Some others are, instead, more or less definite descriptions that express, in addition to that, our approval (*resp.* disapproval) of a presented state of affairs, e.g., 'That has been a really democratic election'. The former are deprived of truth-values, the latter are either true or false. Now, evaluation in science — quite typical in the science of man — is perfectly admissible. On the condition, however, that it leads to a value-judgment of the second kind and that such a judgment is justified in a standard manner, that is, in the same way as all non-evaluative scientific theorems are. In our example, to justify the assessment 'That has been a really democratic election' means to argue that there have been competing political parties, freely and publicly presenting their programs to the potential voters, that an act of voting could be carried out secretly, etc.

In particular, idealization in the humanities is quite often an evaluation. And that is perfectly correct provided that the evaluative idealizational statements of whatever kind (on democracy, free market, capitalist exploitation, etc.) are treated as if they were purely descriptive. That is, as if the only means employed to support them were arguments revealing the appearance of appropriate states of affairs and not persuasive means whose aim was to evoke a positive (*resp.* negative) attitude towards those states of affairs. Indeed, there is no sharp distinction between description and evaluation.

Why, then, the view that there is one is so popular that also the Author seems to follow it? The main reason seems to be the fact of obscurity of the evaluative discourse. Indeed, if we do not know what it means to be a just society, then we can easily complain about the hardly understandable claim that 'This is a just (*resp.* unjust) society' in a way philosophers usually do: it is neither true nor false, hence it is no proposition at all, etc. But the same may be said about the (descriptive) claim 'This is a socratic society', because we cannot know what its meaning could possibly be. Once the definition of it is established (say, ideally socratic society is one in which, for every two members of it, X and Y, for every p, X is inclined to ask Y "p or not p?" and to demand a possibly ramified argumentation for either of the alternatives) such a claim may be more or less easily decided. Quite the same may be said about an evaluation 'S is a just society'. As long as the term is taken to mean what it means in the colloquial language, the meaning of the claim is indefinite and the evaluation seems to have no truth-value. Let us, however, explicate the term 'just': an ideally just society is, say, such a society that for every two members of it, X and Y, and for any social value ν, X and Y have the same initial chance to achieve ν. Now, if we know what is meant by 'just', then 'S is a just society' is true or false, depending on how many social values are more easily attainable in it by privileged people and how large is an appropriate distance between these people and the rest of society. The (pragmatically relevant) fact that such a claim expresses (or evokes, etc.) positive feelings is not to be mixed with the (semantically relevant) fact, whether there is in society S a category of people having a privileged access to a large class of social values, or not.

These remarks intend merely to question the sharpness of the Humean distinction adopted by the Author in the realm of idealization, not the distinction in itself. And the very fact that our language employs the same term 'ideal' both in case of descriptive and prescriptive ideals seems to be an additional confirmation of the above line of reasoning.

REFERENCES

Brzeziński, J. and L. Nowak (Eds) (1992). *Idealization III: Approximation and Truth* (*Poznań Studies in the Philosophy of the Sciences and the Humanities* 25). Amsterdam/Atlanta, GA: Rodopi.

Nowak, L. (1980). *The Structure of Idealization*. Dordrecht/Boston: Reidel.

Nowak, L. (1992). The Idealizational Approach to Science: A Survey. In Brzeziński and Nowak (1992). pp. 9–63.

Leszek Nowak
Uniwersytet im. A. Mickiewicza
Instytut Filozofii
ul. Szamarzewskiego 89 C
60-569 Poznań, Poland
e-mail: epistemo@hum.amu.edu.pl

Lance's Reply to Nowak

Prof. Nowak notes, regarding the points I make about idealization, that I do not "pretend to explain them in terms of the general theory of idealization" (p. 130-131). This is correct. In point of fact, I am skeptical of the very possibility of an adequate and significant theory of idealization and certainly know of none which I endorse. Thus, my remarks should be taken for no more than they are: criticisms of particular idealizations within Bayesian epistemology and general cautionary lessons drawn from these. Since I have not read Prof. Nowak's monograph on idealization, I am in no position to comment on it, but if, as he says, there is a general theory which entails the lessons of the sort I wish to draw, then the theory is to be commended and studied.

I found Prof. Nowak's discussion of the description/prescription distinction a bit puzzling. I do not apply "the standard distinction," whatever that is, nor do I argue that there is in general a "sharp distinction" between descriptive and prescriptive discourse. Even less do I intend to disparage evaluative discourse. Indeed, though I'm not inclined to follow the sort of analysis of normative discourse offered in Prof. Nowak's comments, I've committed myself to the importance of evaluative discourse in the strongest terms, and offered a detailed discussion of its role and significance (1998).

All that I rely on in the paper we are discussing is the distinction between saying that a theory of some normative notion — justification, rationality, moral value — is ideal in the descriptive sense in which it is a simplification of (normative) reality which nonetheless relates to the real situation in some explanatorily significant way, and saying that it is ideal in the sense of accurately describing a normatively ideal entity — perfect rationality, moral perfection, etc. An example is the distinction between an idealized description of real houses, and a description of an ideal house. I presume Prof. Nowak would not deny the importance of keeping these separate.

REFERENCES

Lance, M. and J. O'Leary (1998). *The Grammar of Meaning: Normativity and Semantic Discourse.* Hawthorne: Cambridge University Press.

*Poznań Studies in the Philosophy
of the Sciences and the Humanities
2000, vol. 71, pp. 136–151*

Robert B. Gardner and Michael Wooten

AN APPLICATION OF BAYES' THEOREM TO POPULATION GENETICS

ABSTRACT. In this paper, we derive several conditional probabilities with direct applications in population genetics. The probability that an n-degree relative shares a trait with a given individual is calculated. An example of the application of these probabilities to some previously published DNA fingerprint data is given in which a χ^2 test is performed to determine the degree of relatedness between two individuals.

1. Introduction

The purpose of this paper is to present the derivation of certain conditional probabilities which have applications in population genetics. We make extensive use of Bayes' Theorem and thereby avoid the use of transition matrices (derivations using transition matrices of some of our results have been performed by Li and Sacks 1954). We discuss the application of the derived conditional probabilities to the problem of determining the degree of relatedness between individuals based on their phenotypes and present an analysis of such a case using previously presented DNA fingerprint data.

A detailed account of the use of DNA fingerprinting in the estimation of relatedness was described by Lynch (1988). He showed that the proportion of shared bands is a poor estimate of relatedness unless the frequencies of the bands are near zero (Lynch 1988, Figure 2). This is not surprising since a band with a high relative frequency would be present in significant numbers of unrelated individuals. To compensate for this problem, it is necessary to make a correction in the probability of shared bands for different band frequencies and degrees of relationship. Calculations relative to these corrections are presented here and the results are applicable to any traits which undergo Mendelian inheritance.

2. Results

If a population is in Hardy-Weinberg equilibrium for a given trait, then certain conditional probabilities concerning the presence or absence of the trait in related individuals can be calculated. For example, if an individual demonstrates a trait, the probability that his or her offspring, sibling, cousin, etc. also shows this trait can be calculated.

Consider a dominant trait with allele frequency p. For an individual X, denote the homozygous dominant state as X_1, the heterozygous state as X_2 and the homozygous recessive state as X_3, then the different states of X have the following probabilities: $P(X_1) = p^2$, $P(X_2) = 2p - 2p^2$, $P(X_3) = 1 - 2p + p^2$. However, with a dominant trait, it is unlikely that a heterozygous and a homozygous dominant individual can be distinguished. If the presence of the trait is denoted as $X-$, then $P(X-) = P(X_1) + P(X_2) = 2p - p^2$.

Now, we can do calculations to determine the probabilities of a dominant trait appearing in an offspring of an individual. Denote the known individual as M and the offspring as D. For this calculation, we will need to consider the other parent of D, say F. First, we calculate the probability of the genotypes of the parents, given the genotype of parent M (so when this is nonzero, it will depend only on F_i). Next, we calculate the probabilities for the different genotypes of D given M_i and F_j. Then the probability of each genotype of D given the genotype of M can be derived. With the values in Table 1, we can calculate the probability of each possible genotype of D given any genotype of M as follows:

$$P(D_1|M_1) = \sum_j P(F_j)P(D_1|M_1 \text{ and } F_j) = p,$$

$$P(D_1|M_2) = \sum_j P(F_j)P(D_1|M_2 \text{ and } F_j) = \frac{p}{2},$$

$$P(D_1|M_3) = \sum_j P(F_j)P(D_1|M_3 \text{ and } F_j) = 0,$$

$$P(D_2|M_1) = \sum_j P(F_j)P(D_2|M_1 \text{ and } F_j) = 1 - p,$$

$$P(D_2|M_2) = \sum_j P(F_j)P(D_2|M_2 \text{ and } F_j) = \frac{1}{2},$$

$$P(D_2|M_3) = \sum_j P(F_j)P(D_2|M_3 \text{ and } F_j) = p,$$

$$P(D_3|M_1) = \sum_j P(F_j)P(D_3|M_1 \text{ and } F_j) = 0,$$

$$P(D_3|M_2) = \sum_j P(F_j)P(D_3|M_2 \text{ and } F_j) = \frac{1-p}{2}, \text{ and}$$

$$P(D_3|M_3) = \sum_j P(F_j)P(D_3|M_3 \text{ and } F_j) = 1 - p.$$

M_i	F_j	$P(D_1|M_i \text{ and } F_j)$	$P(D_2|M_i \text{ and } F_j)$	$P(D_3|M_i \text{ and } F_j)$
M_1	F_1	1	0	0
M_1	F_2	1/2	1/2	0
M_1	F_3	0	1	0
M_2	F_1	1/2	1/2	0
M_2	F_2	1/4	1/2	1/4
M_2	F_3	0	1/2	1/2
M_3	F_1	0	1	0
M_3	F_2	0	1/2	1/2
M_3	F_3	0	0	1

Table 1. Conditional Probability of Genotype of Offspring D given the Genotypes of the Parents M and F.

Because we are only concerned with the presence or absence of the trait, the following are obtained:

$$
\begin{aligned}
P(D-|M-) &= P((D_1 \text{ or } D_2)|(M_1 \text{ or } M_2)) \\
&= \frac{P((D_1 \text{ or } D_2) \text{ and } (M_1 \text{ or } M_2))}{P(M_1 \text{ or } M_2)} \\
&= P((D_1 \text{ and } M_1) \text{ or } (D_1 \text{ and } M_2) \text{ or } (D_2 \text{ and } M_1) \text{ or} \\
&\quad (D_2 \text{ and } M_2))/P(M_1 \text{ or } M_2) \\
&= \{P(M_1)[P(D_1|M_1) + P(D_2|M_1)] + P(M_2)[P(D_1|X_2) \\
&\quad +P(D_2|M_2)]\}/(P(M_1) + P(M_2)) \\
&= \frac{1+p-p^2}{2-p}, \\
P(D-|M_3) &= P(D_1 \text{ or } D_2|M_3)) = P(D_1|M_3) + P(D_2|M_3) = p, \\
P(D_3|M-) &= \frac{P(M_1)P(D_3|M_1) + P(M_2)P(D_3|M_2)}{P(M_1) + P(M_2)} = \frac{1-2p+p^2}{2-p},
\end{aligned}
$$

$$P(D_3|M_3) \;=\; 1-p.$$

Now, consider the probability of the presence of a trait in a parent M given the presence of the trait in the offspring D. In this case, also, we have to consider the other parent, F. The calculations are similar to those above and the results are the same, establishing a parent/offspring symmetry. The calculations require the entries of Table 2.

M_i	F_j	$P(M_i \text{ and } F_j|D_1)$	$P(M_i \text{ and } F_j|D_2)$	$P(M_i \text{ and } F_j|D_3)$
M_1	F_1	p^2	0	0
M_1	F_2	$p(1-p)$	$p^2/2$	0
M_1	F_3	0	$p(1-p)/2$	0
M_2	F_1	$p(1-p)$	$p^2/2$	0
M_2	F_2	$(1-p)^2$	$p(1-p)$	p^2
M_2	F_3	0	$(1-p)^2/2$	$p(1-p)$
M_3	F_1	0	$p(1-p)/2$	0
M_3	F_2	0	$(1-p)^2/2$	$p(1-p)$
M_3	F_3	0	0	$(1-p)^2$

Table 2. Conditional Probabilities of Parental Genotypes, M_i and F_j, given the Genotype of an Offspring, D.

For the calculation of the same types of probabilities for two individuals that are siblings, say S^1 and S^2, we use the values from Tables 1 and 2 to get:

$$P(S_1^1|S_1^2) \;=\; \sum_{i,j} P(M_i \text{ and } F_j|S_1^2)P(S_1^1|M_i \text{ and } F_j) = \frac{1+2p+p^2}{4},$$

$$P(S_1^1|S_2^2) \;=\; \sum_{i,j} P(M_i \text{ and } F_j|S_2^2)P(S_1^1|M_i \text{ and } F_j) = \frac{p+p^2}{4},$$

$$P(S_1^1|S_3^2) \;=\; \sum_{i,j} P(M_i \text{ and } F_j|S_3^2)P(S_1^1|M_i \text{ and } F_j) = \frac{p^2}{4},$$

$$P(S_2^1|S_1^2) \;=\; \sum_{i,j} P(M_i \text{ and } F_j|S_1^2)P(S_2^1|M_i \text{ and } F_j) = \frac{1-p^2}{2},$$

$$P(S_2^1|S_2^2) \;=\; \sum_{i,j} P(M_i \text{ and } F_j|S_2^2)P(S_2^1|M_i \text{ and } F_j) = \frac{1+p-p^2}{2},$$

$$P(S_2^1|S_3^2) = \sum_{i,j} P(M_i \text{ and } F_j|S_3^2)P(S_2^1|M_i \text{ and } F_j) = \frac{2p - p^2}{2},$$

$$P(S_3^1|S_1^2) = \sum_{i,j} P(M_i \text{ and } F_j|S_1^2)P(S_3^1|M_i \text{ and } F_j) = \frac{1 - 2p + p^2}{4},$$

$$P(S_3^1|S_2^2) = \sum_{i,j} P(M_i \text{ and } F_j|S_2^2)P(S_3^1|M_i \text{ and } F_j) = \frac{2 - 3p + p^2}{4},$$

$$P(S_3^1|S_3^2) = \sum_{i,j} P(M_i \text{ and } F_j|S_3^2)P(S_3^1|M_i \text{ and } F_j) = \frac{4 - 4p + p^2}{4}.$$

And so for siblings:

$$P(S^1 - |S^2-) = \{P(S_1^2)[P(S_1^1|S_1^2) + P(S_2^1|S_1^2)] + P(S_2^2)[P(S_1^1|S_2^2) +$$
$$P(S_2^1|S_2^2)]\}/(P(S_1^2) + P(S_2^2))$$

$$= \frac{4 + 5p - 6p^2 + p^3}{4(2 - p)},$$

$$P(S_3^1|S^2-) = \frac{P(S_1^2)P(S_3^1|S_1^2) + P(S_2^2)P(S_3^1|S_2^2)}{P(S_1^2) + P(S_2^2)}$$

$$= \frac{4 - 9p + 6p^2 - p^3}{4(2 - p)},$$

$$P(S^1 - |S_3^2) = P(S_1^1|S_3^2) + P(S_2^1|S_3^2) = \frac{4p - p^2}{4},$$

$$P(S_3^1|S_3^2) = \frac{4 - 4p + p^2}{4}.$$

In fact, $P(S^1 - |S^2)$ has already appeared in the literature (Jeffreys et al. 1985).

If two (non-inbreed) individuals are related, then they will be related by a series of offspring to parent steps followed by a series of parent to offspring steps. There may also be a sibling step at the apex of this path. If the number of such steps is n, then these are called n-degree relatives. We show that the calculations of the probabilities allow a commuting of the sibling step with the parent/offspring steps by showing that the numbers are the same for a niece/nephew as they are for an aunt/uncle. So a combination of a parent/offspring step and a sibling step is needed. This corresponds to comparing, for example, a niece, N, and an aunt, A.

Again, comparisons between these two individuals are symmetric and one gets

$$P(A-|N-) \quad = \quad P(N-|A-) = \frac{1+5p-5p^2+p^3}{2(2-p)},$$

$$P(A_3|N-) \quad = \quad P(N_3|A-) = \frac{3-7p+5p^2-p^3}{2(2-p)},$$

$$P(A-|N_3) \quad = \quad P(N-|A_3) = \frac{3p-p^2}{2},$$

$$P(A_3|N_3) \quad = \quad P(N_3|A_3) = \frac{2-3p+p^2}{2}.$$

Calculation of a grandparent/grandchild pair yields the same results.

Surprisingly, even though the probabilities for parents and offspring were different from those of siblings, the grandparent/grandchild probabilities are the same as the aunt/uncle and niece/nephew probabilities. So, if we know the relationship between two individuals (provided they are not siblings), then we can calculate the relevant probabilities by simply using a series of parent/offspring steps, where the sibling step, if present, counts the same as a parent/offspring step.

Now, consider an individual, G^n, that is an n-degree relative of X (e.g. if $n = 1$ then G^1 is a parent, sibling or offspring; if $n = 2$ then G^2 is a grandparent, a grandchild, an aunt, a niece, etc.). Then the following probabilities are derived, which have been established for $n = 1$ (with the relationship "sibling" being a special case) and can be established in general by induction.

$$P(G_1^n|X_1) \quad = \quad \frac{p+(2^{n-1}-1)p^2}{2^{n-1}},$$

$$P(G_2^n|X_1) \quad = \quad \frac{1+(2^n-3)p-(2^n-2)p^2}{2^{n-1}},$$

$$P(G_3^n|X_1) \quad = \quad \frac{(2^{n-1}-1)-(2^n-2)p+(2^{n-1}-1)p^2}{2^{n-1}},$$

$$P(G_1^n|X_2) \quad = \quad \frac{p+(2^n-2)p^2}{2^n},$$

$$P(G_2^n|X_2) \quad = \quad \frac{1+(2^{n+1}-4)p-(2^{n+1}-4)p^2}{2^n},$$

$$P(G_3^n|X_2) \quad = \quad \frac{(2^n-1)-(2^{n+1}-3)p+(2^n-2)p^2}{2^n},$$

$$P(G_1^n|X_3) = \frac{2^{n-1}-1}{2^{n-1}}p^2,$$

$$P(G_2^n|X_3) = \frac{(2^n-1)p-(2^n-2)p^2}{2^{n-1}},$$

$$P(G_3^n|X_3) = \frac{2^{n-1}-(2^n-1)p+(2^{n-1}-1)p^2}{2^{n-1}}.$$

From which:

$$
\begin{aligned}
P(G^n - |X-) &= \{P(X_1)[P(G_1^n|X_1)+P(G_2^n|X_1)] + P(X_2)[P(G_1^n|X_2) \\
&\quad +P(G_2^n|X_2)]\}/[P(X_1)+P(X_2)] \\
&= \frac{1+(2^{n+1}-3)p-(2^{n+1}-3)p^2+(2^{n-1}-1)p^3}{2^{n-1}(2-p)}, \quad (1)
\end{aligned}
$$

$$
\begin{aligned}
P(G^n - |X_3) &= P(G_1^n|X_3)+P(G_2^n|X_3) \\
&= \frac{(2^n-1)p-(2^{n-1}-1)p^2}{2^{n-1}}, \quad (2)
\end{aligned}
$$

$$
\begin{aligned}
P(G_3^n|X-) &= \frac{P(X_1)P(G_3^n|X_1)+P(X_2)p(G_3^n|X_2)}{P(X_1)+P(X_2)} \\
&= \{(2^n-1)-(5\times 2^{n-1}-3)p+(2^{n+1}-3)p^2- \\
&\quad (2^{n-1}-1)p^3\}/(2^{n-1}(2-p)),
\end{aligned}
$$

$$P(G_3^n|X_3) = \frac{2^{n-1}-(2^n-1)p+(2^{n-1}-1)p^2}{2^{n-1}}.$$

Additional observations can be made about these formulae. Notice that if a limit as n approaches infinity is taken, in each of the above formulae, the probability of the presence or absence of the trait is:

$$
\begin{aligned}
P(G^\infty - |X-) &= 2p-p^2, \\
P(G^\infty - |X_3) &= 2p-p^2, \\
P(G_3^\infty|X-) &= (1-p)^2, \\
P(G_3^\infty|X_3) &= (1-p)^2.
\end{aligned}
$$

This is expected since one assumes mating with random unrelated individuals. Notice that if we take $p=0$, then $P(G^n - |X-) = \left(\frac{1}{2}\right)^n$ which is simply the coefficient of relationship (Lynch 1988; Wright 1922).

S_i^2	$P(S_j^2|S_1^1)$	$P(S_j^2|S_2^1)$	$P(S_j^2|S_3^1)$
S_1^2	$(1 + 2p + p^2)/4$	$(p + p^2)/4$	$p^2/4$
S_2^2	$(1 - p^2)/2$	$(1 + p - p^2)/2$	$(2p - p^2)/2$
S_3^2	$(1 - 2p + p^2)/4$	$(2 - 3p + p^2)/4$	$(4 - 4p + p^2)/4$

Table 3. Conditional Probability of the Genotype of an Individual, S_j^2, given the Genotype of a sibling, S^1.

Now consider certain special cases where inbreeding is present. If two siblings S^1 and S^2 have an offspring D then by referring to Tables 1 and 3 (replacing M and F by S^1 and S^2 in Table 1) we get:

$$P(D_1|S_1^1) = \sum_j P(S_j^2|S_1^1)P(D_1|S_1^1 \text{ and } S_j^2) = \frac{1+p}{2},$$

$$P(D_1|S_2^1) = \sum_j P(S_j^2|S_2^1)P(D_1|S_2^1 \text{ and } S_j^2) = \frac{1+2p}{8},$$

$$P(D_1|S_3^1) = \sum_j P(S_j^2|S_3^1)P(D_1|S_3^1 \text{ and } S_j^2) = 0,$$

$$P(D_2|S_1^1) = \sum_j P(S_j^2|S_1^1)P(D_2|S_1^1 \text{ and } S_j^2) = \frac{1-p}{2},$$

$$P(D_2|S_2^1) = \sum_j P(S_j^2|S_2^1)P(D_2|S_2^1 \text{ and } S_j^2) = \frac{1}{2},$$

$$P(D_2|S_3^1) = \sum_j P(S_j^2|S_3^1)P(D_2|S_3^1 \text{ and } S_j^2) = \frac{p}{2},$$

$$P(D_3|S_1^1) = \sum_j P(S_j^2|S_1^1)P(D_3|S_1^1 \text{ and } S_j^2) = 0,$$

$$P(D_3|S_2^1) = \sum_j P(S_j^2|S_2^1)P(D_3|S_2^1 \text{ and } S_j^2) = \frac{3-2p}{8}, \text{ and}$$

$$P(D_3|S_3^1) = \sum_j P(S_j^2|S_3^1)P(D_3|S_3^1 \text{ and } S_j^2) = \frac{2-p}{2}.$$

From which:

$$P(D - |S^1 -) = \{P(S_1^1)[P(D_1|S_1^1) + P(D_2|S_1^1)] + P(S_2^1)[P(D_1|S_2^1)+ \\ \{P(D_2|S_2^1)]\}/(P(S_1^1) + P(S_2^1))$$

$$= \frac{5 + p - 2p^2}{4(2 - p)},$$

$$P(D - |S_3^1) = P(D_1|S_3^1) + P(D_2|S_3^1) = \frac{p}{2},$$

$$P(D_3|S^1 -) = \frac{P(S_1^1)P(D_3|S_1^1) + P(S_2^1)P(D_3|S_2^1)}{P(S_1^1) + P(S_2^1)} = \frac{3 - 5p + 2p^2}{4(2 - p)},$$

$$P(D_3|S_3^1) = \frac{2 - p}{2}.$$

Now consider a similar case, where an individual X mates with an n-degree relative G^n producing an offspring D. See Tables 1 and 4 (replacing M and F by X and G^n in Table 1). These follow:

$$P(D_1|X_1) = \sum_j P(G_j^n|X_1)P(D_1|X_1 \text{ and } G_j^n) = \frac{1 + (2^n - 1)p}{2^n},$$

$$P(D_1|X_2) = \sum_j P(G_j^n|X_2)P(D_1|X_2 \text{ and } G_j^n) = \frac{1 + (2^{n+1} - 2)p}{2^{n+2}},$$

$$P(D_1|X_3) = \sum_j P(G_j^n|X_3)P(D_1|X_3 \text{ and } G_j^n) = 0,$$

$$P(D_2|X_1) = \sum_j P(G_j^n|X_1)P(D_2|X_1 \text{ and } G_j^n) = \frac{(2^n - 1) - (2^n - 1)p}{2^n},$$

$$P(D_2|X_2) = \sum_j P(G_j^n|X_2)P(D_2|X_2 \text{ and } G_j^n) = \frac{1}{2},$$

$$P(D_2|X_3) = \sum_j P(G_j^n|X_3)P(D_2|X_3 \text{ and } G_j^n) = \frac{(2^n - 1)p}{2^n},$$

$$P(D_3|X_1) = \sum_j P(G_j^n|X_1)P(D_3|X_1 \text{ and } G_j^n) = 0,$$

$$P(D_3|X_2) = \sum_j P(G_j^n|X_2)P(D_3|X_2 \text{ and } G_j^n)$$

$$= \frac{(2^{n+1} - 1) - (2^{n+1} - 2)p}{2^{n+2}}, \text{ and}$$

$$P(D_3|X_3) = \sum_j P(G_j^n|X_3)P(D_3|X_3 \text{ and } G_j^n) = \frac{2^n - (2^n - 1)p}{2^n}.$$

G_j^n	$P(G_j^n\|X_1)$	$P(G_j^n\|X_2)$	$P(G_j^n\|X_3)$
G_1^n	$\dfrac{p + (2^{n-1}-1)p^2}{2^{n-1}}$	$\dfrac{p + (2^n-2)p^2}{2^n}$	$\dfrac{(2^{n-1}-1)p^2}{2^{n-1}}$
G_2^n	$\dfrac{1 + (2^n-3)p - (2^n-2)p^2}{2^{n-1}}$	$\dfrac{1 + (2^{n+1}-4)p - (2^{n+1}-4)p^2}{2^n}$	$\dfrac{(2^n-1)p - (2^n-2)p^2}{2^{n-1}}$
G_3^n	$\dfrac{(2^{n-1}-1) - (2^n-2)p + (2^{n-1}-1)p^2}{2^{n-1}}$	$\dfrac{(2^n-1) - (2^{n+1}-3)p + (2^n-2)p^2}{2^n}$	$\dfrac{2^{n-1} - (2^n-1)p + (2^{n-1}-1)p^2}{2^{n-1}}$

Table 4. Conditional Probability of Genotype of n-degree Relatives, X and G^n, given the Genotype of X.

And hence:

$$
\begin{aligned}
P(D-|X-) &= \{P(X_1)[P(D_1|X_1)+P(D_2|X_1)]+P(X_2)[P(D_1|X_2)+ \\
&\quad P(D_2|X_2)]\}/(P(X_1)+P(X_2)) \\
&= \frac{(2^{n+1}+1)+(2^{n+1}-3)p-(2^{n+1}-2)p^2}{2^{n+1}(2-p)}, \\
P(D-|X_3) &= P(D_1|X_3)+P(D_2|X_3) = \frac{(2^n-1)p}{2^n}, \\
P(D_3|X-) &= \frac{P(X_1)P(D_3|X_1)+P(X_2)P(D_3|X_2)}{P(X_1)+P(X_2)} \\
&= \frac{(2^{n+1}-1)-(2^{n+2}-3)p+(2^{n+1}-2)p^2}{2^{n+1}(2-p)}, \\
P(D_3|X_3) &= \frac{2^n-(2^n-1)p}{2^n}.
\end{aligned}
$$

Notice that for $n=1$, this also reduces to the case of mated siblings. Also, for any particular type of inbreeding, say continued brother/sister mating, it is possible to calculate these type probabilities. The calculations (although tedious) would employ the methods used here, that is, the construction of tables of the nine possible genotypes of the parents with the corresponding conditional probabilities.

3. An Application

Jeffreys *et al.* (1985) presented the following example of the use of DNA fingerprints in determining relationships: "The case concerned a Ghanaian boy born in the United Kingdom who emigrated to Ghana to join his father and subsequently returned alone to the United Kingdom to be reunited with his mother, brother and two sisters. However, there was evidence to suggest that a substitution might have occurred, either for an unrelated boy, or a son of a sister of the mother... As a result, the returning boy was not granted residence in the United Kingdom." Conventional genetic markers indicated that the woman and boy were related (with 99% probability), but could not determine whether the woman was the boy's mother or aunt. DNA fingerprints were produced from blood DNA samples taken from the boy, the mother, her three other children, and an unrelated individual. The father was unavailable. Based on the probability of unrelated individuals sharing a band, the allele frequency for a band was calculated to be $p = 0.14$ (all bands were assumed to have the same frequency). The mother and boy were found to share 25 maternal specific bands. Based on this, it was calculated that the probability of these two being unrelated was $(0.26)^{25} = 2 \times 10^{-15}$ (an allele with frequency 0.14 will appear in an individual with probability $2(0.14) - (0.14)^2 = 0.26$). The corresponding probability of the mother actually being the aunt of the boy was said to be 6×10^{-6}. This latter calculation, however, was found to be erroneous and declared irrelevant by Hill (1986) who approached this same problem from a maximum likelihood viewpoint, and reached the same conclusion. The boy was granted residence in the United Kingdom. The method used by Jeffreys *et al.* (1985) is very restricted and not suitable for general use in testing specified relationships.

We can re-analyze this case using the established probabilities. It was remarked that there are two relationships to be tested, that of aunt/nephew ($n = 2$) and mother/son ($n = 1$). Using equations (1) and (2) with $p = 0.14$ for all bands, as in the paper, we can generate expected numbers of shared bands and do a χ^2 test on these two relationships and on the possibility of no relation between the mother and boy. We get the results of Table 5 and find that, of these three possibilities, we fail to reject only the option of a mother/son relationship.

Null Hypothesis	χ^2	Conclusion
M and X are unrelated	49.4	reject
M and X are aunt/nephew	12.5	reject
M and X are mother/son	.75	fail to reject

Table 5. χ^2 Values for the Three Plausible Null Hypotheses for the Information of Jeffreys *et al.* (1985). M is the Alleged Mother and X the Alleged Son.

4. Conclusion

In conclusion, we have presented conditional probabilities on individuals sharing traits which undergo Mendelian inheritance given that the degree of relatedness of the individuals is known. An application of these conditional probabilities to the estimation of degree of relatedness using a χ^2 test has been given. An alternative approach to estimating relatedness would be to calculate likelihoods of certain degrees of relationships, given data sets. It is our belief that these methods would be valid and unambiguous in establishing close degrees of relatedness ($n = 1$ or $n = 2$) and when average allele frequencies are low (say 15% or less). However, these procedures may become inaccurate when applied to more distant relationships or when average allele frequencies are high (Lewin 1989; Lynch 1988). If the phenotypic data of several individuals chosen from a single population are available, this problem could possibly be overcome by piecing together a pedigree using only the close relationships. There can be a high degree of confidence in these close relationships, and therefore a fairly high degree of confidence in the derived pedigree.

The accurate derivation of pedigrees would be especially important in selective breeding programs in small populations or endangered species. Also, inbreeding could be controlled in such a situation if relatedness between individuals were known.

Acknowledgements. Support for this research was provided by the Alabama Agricultural Experiment Station (AAES no. 15-913078). Comments by S. Dobson, C. Guyer and G. Hepp on an earlier draft of this manuscript are gratefully acknowledged.

REFERENCES

Hill, W. (1986). DNA Fingerprint Analysis in Immigration Test-Case. *Nature* **322**, 290–291.

Jeffreys, A., F. Brookfield, and R. Semeonoff (1985). Positive Identification of an Immigration Test-Case using Human DNA Fingerprints. *Nature* **317**, 818–819.

Lewin, R. (1989). Limits to DNA Fingerprinting. *Science* **243**, 1549–1551.

Li, C. and L. Sacks (1954). The Derivation of Joint distribution and Correlation between Relatives by the use of Stochastic Matrices. *Biometrics* **10**, 347–360.

Lynch, M. (1988). Estimation of Relatedness by DNA Fingerprinting. *Molecular Biology and Evolution* **5**(5), 584–599.

Wright, S. (1922). Coefficients of Inbreeding and Relationship. *American Naturalist* **56**, 330–338.

Robert Gardner
Department of Mathematics
Department of Physics and Astro.
East Tennessee State University
Johnson City, Tennessee 37614
e-mail: gardnerr@etsu.edu

Michael Wooten
Department of Zoology
and Wildlife Science
AL Ag. Experiment Station
Auburn University
Auburn, Alabama 36830-5414
e-mail:mwooten@ag.auburn.edu

Commentary by Lynne Seymour

Bayes' Theorem is an elementary probabilistic mechanism with such broad implications that an entire branch of statistical inference has arisen from it. The authors have used this mechanism to simplify some basic probability calculations in population genetics. Why wasn't this done in the first place? Almost surely because the scientist who first derived those probabilities was unaware of the tools which could simplify his work.

This is a common occurrence at the interface of mathematics with the sciences. In particular, it happens in statistics, albeit to a slightly lesser extent since scientists tend to use statistics in their everyday research. Even so, few scientists truly understand the statistical procedures they use — procedures on which they base the conclusions of their livelihoods! The χ^2 test that the authors used will be employed here to highlight the logic inherent in a test of hypotheses, and to underscore how probability can be invaluable in interpreting the hypothesis test.

First, what is a statistical test of hypotheses? In general, it is a logical procedure which is derived under stated conditions. Null and alternative hypotheses are clearly stated. A random variable — a function of data called the *test statistic* — is developed and its distribution is derived and/or approximated assuming that the null hypothesis is true. A value for the test statistic is calculated using the actual data, and the likelihood of seeing this value from the test statistic's distribution under the null hypothesis is evaluated. If this value is judged to be unlikely under the null hypothesis, then the null hypothesis is rejected at an explicitly stated *significance level*. This significance level, usually taken to be 5%, is interpreted as the largest chance one is willing to tolerate of wrongly rejecting the null hypothesis. If any of the assumptions under which the test is derived are violated, the results of the test cannot be trusted.

Many common statistical tests have "shorthand" names by which they are invoked. The authors, in applying their newly-derived probabilities to an already-published example, state that they "do a χ^2 test" to draw a conclusion. Which χ^2 test would that be? A statistician knows that there are a great many of them. An elementary text which assumes only a modest knowledge of high school algebra (Anderson *et al.* 1994) lists three χ^2 tests in the index. One is for the variance of a single population (which the reader can rule out fairly quickly); another is for independence (which, upon close inspection of the author's Table 5, may be ruled out); and yet another is for goodness of fit (which is the one used).

Consider the first of the χ^2 *goodness-of-fit* tests which the authors used. The null hypothesis is $H_0 : M$ *and* X *are unrelated*, and the alternative is $H_1 : M$ *and* X *are related*. The data indicates whether a matched pair

of bands is the same or different. The test statistic requires that there
be k categories into which each observation falls exactly once. Hence the
$k = 2$ categories may be labeled *same bands* and *different bands*. The test
statistic is given by $\sum_{i=1}^{k} \frac{(O_i - E_i)^2}{E_i}$, where O_i is the observed number of
bands in the ith category, computed assuming H_0 is true. Under H_0, this
test statistic has a χ^2 distribution with $k - 1 = 1$ degree(s) of freedom,
denoted $\chi^2(1)$. (Note that Table 5 does not specify the degrees of freedom
for the χ^2 distribution.) From the $\chi^2(1)$ table, it becomes apparent that
the conclusions given in Table 5 are not only at the 5% significance level,
but certainly at any reasonable significance level.

Most statistical hypothesis tests today cite a *p-value* rather than a
significance level (in fact, statistical analysis packages such as SAS report
only *p*-values when performing statistical tests of hypotheses). The *p*-
value is the tail probability associated with the test statistic: *i.e.*, it is the
conditional probability of obtaining a test statistic more in line with the
alternative hypothesis than the one observed, given that H_0 is true. The
traditional significance level is the largest such probability one is willing
to tolerate when rejecting H_0.

The *p*-values associated with the $\chi^2(1)$ statistics given in Table 5 are
summarized below (they were calculated using the function CHIDIST in
Excel). These *p*-values add an exclamation point to the example which
cannot be communicated by the raw χ^2 statistics. Roughly, based on the
evidence presented, they say: If one rejects that M and X are unrelated,
the chance of being wrong is negligible; if one rejects that M and X are
aunt/nephew, there is a .04% chance of being wrong; and if one rejects
that M and X are mother/son, there is a 38.64% chance of being wrong.

Null Hypothesis	$\chi^2(1)$	*p*-value
M and X are unrelated	49.40	2.09×10^{-12}
M and X are aunt/nephew	12.50	0.000407
M and X are mother/son	0.75	0.386476

REFERENCES

Anderson, D., D. Sweeney, and T. Williams (1994). *Introduction to Statistics: Con-
    cepts and Applications* 3rd ed. St. Paul: West Publishing.

Lynne Seymour
Department of Statistics
University of Georgia
Athens, GA 30602-1952
e-mail: seymour@stat.uga.edu

Gardner and Wooten's Reply to Seymour

We wish to thank Dr. Seymour for her commentary. One of us (Gardner) is a pure mathematician, while the other (Wooten) is a population geneticist. Our grasp of the subtleties of statistics is, no doubt, not optimal. We do enthusiastically agree, however, that probability is "invaluable" in any statistical process.

We also agree that the assumptions of a statistical test must not be violated (of course). We should probably emphasize that our χ^2 test of Section 3 assumes that the relevant bands are inherited independently (that is, there is no linkage between the bands). We think this a reasonable assumption and note that the two previous analyzes of this case (by Jeffreys *et al.* and Hill) have also (implicitly) assumed no linkage between the bands.

The use of p-values, as mentioned by Dr. Seymour, is certainly preferable to our "significance level" approach (we chose our approach due to our limited access to the appropriate software). We thank her for the "exclamation point"!

In conclusion, we mention that an additional problem with our particular approach to the application of the derived probabilities lies in the area of variance. If there is a large amount of variance within a population, then the recognition of distant relatives will be difficult due to the "noise" produced by the variance. This is addressed in the Lynch paper referenced in our article.

Poznań Studies in the Philosophy of the Sciences and the Humanities
2000, *vol.* 71, *pp.* 152–167

Peter D. Johnson, Jr.

ANOTHER LOOK AT GROUP SELECTION

ABSTRACT. Group selection is currently widely reviled as an explanatory device in the theory of evolution. As a counter to this revulsion, here it is proposed that the existence of the eusocial communal "superorganisms," composed of multicellular organisms highly differentiated according to their function within the group, constitutes strong evidence that selection at the group level has indeed played a significant role in animal evolution. Furthermore, it is proposed that the development of the eusocials is the fourth and most recent "great leap" of organization of life, in each of which leaps the new form of organism is a cooperative, coadapted "group" of individuals from the previous highest level of organization. In each of these leaps, the rôle of group selection seems indispensable. The anti-group-selectionist view that symbiotic coadaptation and cooperation within a group must be explained purely in terms of the relative advantage to the individuals involved is addressed. An example from the human sphere is presented.

About 20 years ago a Thought occurred to me that has undoubtedly occurred in various garb to hundreds, if not thousands, of amateur biologists over the past century: that the organized groups of eusocial insects, the hives and nests of bees and (some) wasps and the colonies of ants and termites, constitute a new variety of organism, of which the constituents are themselves multicellular organisms, differentiated and adapted to their function within the group in faint but unmistakable analogy to the differentiation and adaptation of cells within multicellular organisms. It may seem a stretch to compare the modest differentiation of the insects into workers, warriors, nurses, queen and drones to the extreme differentiation of cells in a multicellular organism, but this objection pales if you allow that this new sort of organism is in its early days — perhaps we should say, early eons. For a fair empirical test of the claimed analogy, we really ought to let the great experiment of evolution run for another half-billion years from now, or for however long we think it took multicellular organisms worthy of the name to evolve from their tentative forerunners.

This ultimate test seems beyond our means at the moment. But there are a couple of items supporting the analogy beyond just the blissful, raw perception of similarity. One is the blissful, raw perception of another similarity, that between the *processes* of differentiation within multicellular organisms and within eusocial insect groups. In the latter, how an individual insect turns out depends, roughly speaking, on "treatments" administered by nurses during larval and pupal development; the nurses are apparently informed what the group needs by some subtle communication system within the group. Is this not reminiscent of the differentiation of stem cells in embryo, with the process regulated by a chemical communication system of messenger hormones and enzyme triggers? The cell differentiation process is the more sophisticated, certainly, but the point is that in each case, the undifferentiated individual awaits marching orders from the group.

The second item is more far-fetched, well up the scale of grand patterns visible only at a great distance. Somebody has to point this out, and it may as well be me, an outsider impervious to ridicule: given the speculative account of the development of life on earth that we now possess, the development of super-organisms composed of co-adapted, highly differentiated multicellular organisms seems the inevitable "next thing" in (animal) evolution. According to that account, pre-biotic organic molecules joined to form large self-replicating molecules; these teamed up somehow to form simple prokaryotic cells; prokaryotes combined to form complicated eukaryotic cells; eukaryotes were assembled (by natural selection) into multicellular organisms. In each case, what we see today, or in the fossil record, are relatively finished products tuned by hundreds of millions or billions of years of evolution. Natural selection will most probably have done away with the early, crude cooperatives of individuals at level $x - 1$, these cooperatives being the forerunners or the early exemplars, depending on your definition, of the creatures of level x. (Some traces from the transitional epochs may persist; but a discussion of these traces would take us too far afield for this essay.) How interesting (I thought to myself, twenty years ago) to be sentient in an epoch during which a leap to the next level is gathering in a corner of the animal kingdom. What a chance to get an idea of how these things work! (Although the shortness of our life spans, as individuals or even as successions of interested observers, severely limits the reliability of our conclusions, which is not necessarily a bad thing. Amateurs, rejoice! This is one area where, it appears, we are forcibly confined by circumstance to the Province of Pleasurable Speculation.)

With what excitement did I run across Honeycutt's (1992) "Naked Mole-Rats"! This is a report on a eusocial species of rodent, found in Africa, the organized groups (colonies) of which exhibit the same general characteristics as the groups of eusocial insects: crude but undeniable differentiation of the individual mole-rats in accord with their function within the group, the differentiation apparently effected by purposeful differences in upbringing; a subtle but effective communication system within the colony; and a queen leader who bears all the young. To my mind, this was another example supporting the argument for the existence of the Great Leap to Level Five of Animal Life, all the more satisfying in that it involves Level Four organisms very far removed from insects.

It was in this article that I first encountered the term "haplodiploidy" (you will conclude that I am poorly read, and you will be right), and this struck a disturbing note. Haplodiploidy describes the genetic conditions within most groups of eusocial insects whereby the drones are haploid, carrying unpaired chromosomes in their cells, as in the sex cells of most multicellular creatures, and the queens are diploid, like most animals — so their haploid ova carry a random selection of representatives of their chromosome pairs. Without going into the arithmetic, the upshot of all this is that sibling offspring of a mating of a queen and a drone are, on average, more closely related to each other than any is to their mother, the queen. This circumstance is adduced to explain the apparently altruistic behavior of the workers, warriors, and nurses (or, just the workers, in the majority of cases) in sacrificing their own chance for reproduction to aid in the care, feeding, and protection of their siblings. Haplodiploidy was broached in Honeycutt's (1992) because the mole rats do not exhibit haplodiploidy, yet the non-reproducers among them behave as altruistically as worker bees, toiling and sacrificing apparently for the good of the group. Rodney Honeycutt duly notes this lack of haplodiploidy among the mole-rats, as well as a symmetric presence of haplodiploidy among numerous non-eusocial insect species, but stops about ten miles short of rejecting haplodiploidy as an explanation of altruistic behavior toward siblings. He lets the non-eusocial haplodiploidy hang as a vague warning, and disposes of altruism among the naked mole-rats by suggesting an explanation similar to haplodiploidy: the suggestion is that because of incest in mole-rat breeding, sibling genetic relations are closer, on the average, than parent-child genetic relations. This genre of explanation falls under the heading of *kin selection*, closely related to the term *inclusive fitness*.

To me, there seemed to be something very wrong with this sort of explanation. If these explanations are sound, and if the eusocial individuals are to their group as the cells of our bodies are to our bodies, then the altruistic behavior of, say, our bone cells, selflessly dedicated to their bony task, apparently for the good of the body, without a chance of participating in reproduction, would be explained by the fact that they are genetically identical to the other cells in the body (except for the occasional mutation). This justification of bone cell behavior seemed to me absurd, and the haplodiploidy justification of eusocial insect behavior no less so. However, I came to realize that these justifications and explanations are regarded as the acme of scientific rigor in some precincts, and that I would need something more than my sense of their absurdity to convince the zealous believers of their error.

The fact that haplodiploidy occurs among insects that are not eusocial can be dismissed as evidence against the inclusive fitness explanation of apparent altruism on the grounds of "non-determinism" or "contingent circumstance," whatever it should be called; the argument is that gene-sharing between siblings makes selection for altruism possible, but does not dictate or determine that altruism will necessarily develop. So scratch that argument. Here is another that seems to me to be logically devastating, but the inclusive fitness mavens will doubtless find a way to ooze around it: in those instances where sibling altruism is supposedly rewarded because the siblings are *more* closely related to each other than to their parents, at least to their mother queen, why are there queens? These queens seem to be at a reproductive disadvantage to their offspring, no? The obvious rejoinder is that without the queen, the whole show falls apart, doesn't exist, and to that I reply — exactly! Gotcha! You cannot explain what goes on in these groups without referring to the actual organization of the group qua group — you cannot explain *everything* by the relative levels of fitness, inclusive and otherwise, of the constituent individuals.

Given the intensity of the anti-group-selectionism currently prevalent, this argument works like an épée against a tank — I touch your fuel line with my épée briefly and suddenly it is ground to confetti by the tank treads. Plus, this argument does not release bone cells from their gene-sharing strategy. So I will not struggle further with this weapon, although I still regard it as perfectly respectable. But no matter; in the course of preparing this essay, I stumbled upon something that orders the argument against gene-sharing explanations of behavior within eusocial groups, as well as of "behavior" of specialized cells within a multicellular organism. Ironically, the two big words that explain it all come from the brilliant Richard Dawkins, one of the most vehement of the anti-group-selectionists, and in particular from one of the added chapters of his 1989 new edition of *The Selfish Gene*: the two words are *bottleneck reproduction*. These will refer here to the situation undebatably observable in multicellular organisms, viewed as groups of cells, and in eusocial groups, whereby reproduction is an activity carried on by a few special, well-adapted individuals, surrounded by other well-adapted individuals that do not participate in reproduction, but provide sustenance, protection, or some other service to the group. It is not only undebatable that reproduction in these groups or bodies is the province of a few individuals, it is also undebatable that, as noted earlier, the non-reproducing individuals in the support teams are *brought up* to be as they are. They are not "free" individuals "choosing" among a variety of reproductive strategies, but rather slaves in a fascist state, instruments of the organism they inhabit. My contention is that the "altruistic behavior" of our bone cells, or of the worker bees, needs no justificatory explanation, because it is not really "behavior," it is "function," in quite a different category from, for instance, the apparently similar behavior exhibited by helper offspring in jackal or coyote families (see Moehlman 1987).

These families seem to be groups teetering on the edge of eusociality and true bottle-neck reproduction, and therefore may provide clues as to how eusocial groups came to be. If some of these families were to develop, by mutation or by a fortunate mix of genes already present in the mating pair, a form of pup-rearing whereby a certain percentage of the pups are brought up to be *nothing but* specialized helpers, with no possibility of becoming members of a mating pair (one thinks of the castrati in Renaissance Europe, for instance, or the eunuch bureaucrats of the Ming dynasty), and if this development proved sufficiently advantageous to the family group that groups with this feature tended to survive and prosper more than groups without it, i.e., if there were selection pressure at the group level for this group trait, then we would have eusocial coyotes or jackals, the helpers would have joined the slave category of bone cells and worker bees, and we could reasonably declare the family groups to be coherent organisms. Anti-group-selectionists will scoff at this rosy scenario, but let me ask them, how has eusociality come about? For that matter, how have multicellular organisms come to exist? Undoubtedly, in any of the evolutionary leaps to new organismic levels, there were numerous failed experiments and close calls taking place over hundreds of millions of years, before organisms at the new level definitely achieved existence. (For those who persist in emphasing the weakness and insignificance of selection at the group level, it should be pointed out that, were you a sentient observer in the era when the prokaryotes were the ruling life forms, the development of eukaryotes would undoubtedly have appeared to you to be an extremely sporadic and curious sideshow to the main event, the evolutionary competition among the teeming prokaryotes. The analogy with our era is obvious).

The case of the helper jackals and coyotes is one in which inclusive fitness and the arithmetic of gene-sharing really are valid tools of analysis. If there is one point on which the anti-group-selectionists are absolutely, indisputably right, it is that an individual trait that benefits the group but diminishes the individual's reproductive fitness will die out. Therefore, an individual trait that does not die out must not diminish the individual's reproductive fitness. In the case of altruistic behavior (in pre-organismic groups that do not yet exhibit bottleneck reproduction) that *apparently* puts the individual at a disadvantage, how could this happen? One obvious possibility is that group evolution has provided a "co-adaptation," a (possibly inobvious) compensation to the individual for their altruism that raises their individual reproductive fitness up to or above the average. In the case of the jackal and coyote helpers, their relatedness to their siblings could well constitute part of the compensation package, a free gift from Mother Nature or Dame Fortune to the group. It is very important to observe that this contention about the effects of gene-sharing, that it makes possible the development of a group trait that may (or may not) increase the fitness of the group, is not at all in contradiction to the view that gene-sharing must be added into the calculation of the individual's fitness. The point that must be made, that seems so obvious that I wonder why we have such difficulty in grasping it, is that selection at the group and individual levels are not in opposition, but rather in superposition. Developments at either level affect the other, but that does not mean that there is some sort of rivalry concerning relative importance between selection at the two levels, any more than there is a rivalry between the earth's orbit around the sun and the moon's orbit around the earth as to which is the more important in determining the moon's orbit around the sun.

No, the rivalry between selection at the group and individual levels is a misunderstanding, an illusion in the minds of the humans, and what a fierce, silly illusion it is. To get an idea of how fierce, it is instructive to read Richard Dawkins' (1994) commentary on Wilson and Sober's (1994) brave and quite reasonable "Reintroducing Group Selection to the Human Behavioral Sciences." "They are zealots, baffled by the failure of the rest of us to agree with them. I can sympathize: I remain reciprocally baffled by what I still see as the sheer wanton, head-in-bag perversity of the position they champion" (1994, p. 616). Dawkins' attitude toward group selection resembles the attitude of a John Birch Society member toward Marxism: he just will not have it and he will take every opportunity to heap scorn upon it, "it" being any attempt to consider selection at group level as a significant force in evolution. How does he deal with the existence of eusocial species, or the view of multicellular organisms as "eusocial" aggregates of individual cells? Take a look at *The Selfish Gene*, 1989 edition, especially the two added chapters, and their end notes. To haplodiploidy, gene-sharing, kin selection, inclusive fitness, and game theory, all useful in explaining everything that needs to be explained in terms of selection at the individual level, Dawkins has added a marvelous new rhetorical invention, the "extended phenotype," a concept whereby the "gene," the useful but possibly mythical entity that is the basis of all life, inhabits and survives in, or not, the whole great universe. This concept is obviously "correct," but the way Dawkins has tarted it up, it can be used to explain *anything*. Dawkins uses it to explain why groups don't exist, sort of. In fact, in his commentary on Wilson and Sober's paper, he is toying with the idea of doing away with multicellular organisms altogether. Somebody stop this man!

Speaking of genes, here is a matter for further consideration: let's consider the view that genes are like quarks and leptons in quantum mechanics — entities whose physical reality is dubious, but they come in devilish useful in explaining phenomena. This is certainly the view that underlies Dawkins' brilliant theory. The problem with the analogy is that genes are commonly thought by practical biological researchers and technicians to have a definite physical incarnation: they are relatively short nucleotide sequences on chromosomes. Now, if there is one thing about consecutive subsequences of sequences that is axiomatic, it is that they are determined by their start and their finish. How does one determine where one gene leaves off and the other begins? The practical biologists have their methods. I want to point out that genes, by their *physical nature*, are embedded in *groups*, in the form of sequences, and that co-adaptation and teamwork, rather than being dubious, wishful hypotheses promulgated by fuzzy-minded old gaffers looking for happy endings may be the very basis of life. But perhaps this is going too far.

One possible explanation for the ferocity of anti-group-selectionism is that the intellectual mistakes of the evolution theorists of the 60's and 70's, not to mention the casual errors of earlier luminaries like Konrad Lorenz, with their "good-of-the-species" toss-offs, engendered such a violent reaction among the next generation that they veered to an erroneous extreme, the doctrine that groups are *never*, under *any circumstances*, to be allowed to be considered in evolutionary theory. The anti-group-selectionists' unreasonable attachment to this doctrine was reinforced, I believe, by the development of the mathematical side of their theory, the gene-sharing arithmetic, the game theory, the calculation of evolutionarily stable strategies, and all that. A little mathematical success can be a dangerous thing — it generates hubris, disdain, and an emotional attachment to one's own way of thinking; it appears that the anti-group-selectionists may have succumbed to the intoxication of self-admiration brought on by mathematical success.

If evolution can be understood without mentioning group selection, that does not mean that it should be so understood. A geocentric view of the universe is possible, but few today would consider it desirable. Analogously, it seems a foolish exercise to adhere to a discipline that forbids reference to what seems to some of us an important part of what we are trying to understand. As I indicated at the beginning of this essay, co-adaptation within groups and the development of different levels of group organization seem to me to be fascinating phenomena that we should be actively studying, rather than avoiding mention or denying the existence of.

Granted, in the superposition of selection at the individual level and selection at the group level, for groups that have not achieved organism status, the group selection component is small compared to the individual selection component. So what? Does that mean that it is not to be considered at all? We keep hearing that group selection will be "swamped" by individual selection. Those who maintain this do not understand what it means for the two sorts of selection to be superposed, rather than opposed. To borrow an analogy from James Abu-Rizk, think of a big ocean liner steaming along, with a little fish attached somehow to its bow, pushing it at a right angle to its path. Is the force supplied by the fish very small compared to that provided by the ship's engines? Yes, very small indeed. Does that mean that the fish will not be able to change the ship's course? No — absent a countervailing force working against the fish, the ship will turn.

My thesis is that the existence of eusocial groups is strong evidence that the little fish of group selection has turned the ocean liner of evolution. This contention brings up another complaint I have to make against the anti-group-selectionists, their attempted dismissal of group selection by mathematical modeling. For an instance of this see Grafen (1984). Mathematical tools can be useful and elegant, but beware their seductive powers! Their very elegance can cloud the mind! Let us never forget that if a mathematical model fails to explain or "disproves" a natural observation, then either the observation is erroneous or the model is defective, as a model of whatever it is a model of. We do not reject reality in favor of the model, however elegant the model.

I hear you about to pounce, anti-group-selectionist! You will claim that the observation of co-adaptation within groups and the development of organismic coherence is not being rejected by all this modeling, but rather the application of the postulate of selection at the group level, as an explanation. We do not need the little fish of group selection, you will say, because our models and calculations can explain everything with reference only to selection at lower levels. I leave it to others to judge how convincing the actual models are, but the claim that group selection should be eschewed because it is *unnecessary* is very interesting. It is really the same objection put to me by somebody at last autumn's (1996) SEP Conference, that group selection violates the principle of Occam's

Razor, "thou shalt not multiply entities unnecessarily." Now, William of Occam was writing at a time when natural philosophers were ascribing mundane phenomena, such as the fall of a rock when dropped, to the activities of invisible angels, and one can see that the Razor comes in handy in cutting out that sort of philosophizing, but nobody really thinks that the unqualified Razor should be elevated to the status of an axiom of meta-science. (See "Occam's disposable razor" 1996, pp. 81–82.) For one thing, strict observance would mean saying good-bye to the periodic table. And I suppose that we would have to eschew reference to multi-cellular organisms, or even cells, in discussing evolution, since everything can, at bottom, be explained by the activities of genes, or their physical incarnations. Perhaps we would be confined to phrases like "the assemblage of DNA, RNA, and a supporting cast of organic molecules formerly referred to as a coyote."

I contend that there are, in nature, clear instances of co-adaptation within groups that are most easily understood as having been brought about by group selection. All attempts to dismiss these instances on the grounds that they do not fit a particular model are invalid. Dismissing these instances by giving cogent accounts of them not involving group selection can be valid — this would entail a convincing argument that selection at the group level has not, in fact, taken place during these developments, or has taken place but has not had any effect on the co-adaptation of interest. I also contend that not enough is known about the nature of co-adaptation to support attempts at mathematical modeling at this point. Humans observed the stars and planets for hundreds of thousands of years before attempting a model of these things. I think we should collect a few more observations (quantitative, if possible) on co-adaptation and organismic coherence before trying to jazz things up with mathematics.

Here is an illustration that I hope will illuminate and support the contentions above. Taken from human experience of humans, it is a much shakier kind of example than the examples of the eusocials, which I consider "strong" examples in support of the efficacy of group selection. Perhaps the controversial weakness of this example will advance the discussion.

The example is the nearly universally remarked generous bellicosity of young men, their willingness to go to war, to get into fights, or to participate in dangerous sports. Here we have a trait *apparently* disadvantageous to the individual, but clearly of possible advantage to the small groups of humans that roamed the earth for a million years or so before the onset of civilization. On the grounds that a trait that does not die out must not be disadvantageous to the individual, it follows that there must be something canceling the apparent disadvantage of this trait. The view I am trying to promulgate is that the role of selection at the group level in the development of this trait, and of the mechanisms of compensation that make the propagation of this trait possible, is not only not unreasonable to allow in the discussion, it would be unreasonable *not to* allow it. The groups in which these things developed have long since dissolved; so what? Does this mean that differences in survivability among these groups had no role in the development of these things? Think not!

I see the anti-group-selectionist argument coming, based on the most obvious of the compensations for this generous bellicosity: girls like it. So the persistence of the trait in the young men can be explained under the same rubric as the peacock's tail; a display of reckless abandon in the males is a display of health, a sign that the offspring of this crazy guy will be bursting with animal spirits. So there is no reason to adduce the effect of group selection in explaining these matters, would claim the anti-group-selectionist. On the contrary, it is quite reasonable to suppose that both the male trait and the compensating female reaction to it benefited from group selection. The simplified scenario is that in groups that did not develop some sort of compensation for the male trait, the trait would have died out, the groups would have been left relatively defenseless, and would have eventually succumbed — on the average, that is to say, with greater frequency than groups enjoying the services of bellicose young men.

For those who would sneer at this scenario, I have the observation of another component of the compensation package for feisty young men: the stigmatization up to ostracism of the young men that do not exhibit enthusiasm for mortal combat. Have you ever heard references to President Clinton as a "draft-dodger"? What about the people that insist on despising Muhammad Ali for his unwillingness to be drafted in the era of Viet Nam? He lost the best years of his life in prison, after one of the most moving pronouncements in history, "I ain't got no fight with them Viet Congs," and later became canonized for his overall strength of character and his contribution to black America and Islam in black America, yet you can still find people, mostly older men, who will not hear good of him. What is it to them? What advantage do they get by this behavior? I maintain that this is a pure effect of group selection; there is no individual advantage or disadvantage to being a disapproving elder, the advantage resides purely at the group level, in the evening-up of the competition between the young men that will go and those that won't.

I hope that you will agree that any attempts to insert these conjectured co-adaptations and compensations into a mathematical model would be premature.

My thanks to Niall Shanks, whose constructive disagreement with all of my views was stimulating, and who referred me to the paper of Wilson and Sober.

REFERENCES

Dawkins, R. (1989). *The Selfish Gene*. Oxford: Oxford University Press.
Dawkins, R. (1994) Burying the Vehicle. *Behavioral Brain Sciences* **17**, 616–617.
Editorial (1996). Occam's Disposable Razor: Is Seeing Believing? *The Economist* Oct. 5–11, 81–82.
Grafen, A. (1984). Natural Selection, Kin Selection, and Group Selection. In J. Krebs and N. Davies (Eds.), *Behavioural Ecology*. Sunderland, Massachusetts: Sinauwer Associates, Inc. pp. 62–84.
Honeycutt, R. (1992). Naked Mole-Rats. *American Scientist* Jan.-Feb., 53–54.
Moehlman, P. (1987). Social Organization in Jackals. *American Scientist* July-August, 366–375.
Wilson, D. and E. Sober (1994). Reintroducing Group Selection to the Human Behavioral Sciences. *Behavioral and Brain Sciences* **17**, 575–654.

Peter D. Johnson, Jr.
Department of Discrete and Statistical Sciences
120 Math Annex
Auburn University, Alabama 36849
e-mail: johnspd@mail.auburn.edu

Commentary by Niall Shanks

Contrary to Johnson's closing remarks, we do have some points of agreement. I agree, for example, that evolutionary processes do not necessarily take the most parsimonious path. I agree that mathematical models can be seductively misleading. I agree that there is probably something conceptually unhealthy about the genic reductionism that has emerged from some species of population biology. I even agree that group selection may be *possible*. But for all this, is there such a thing as *group selection*?

The biological hierarchy of organization runs from macromolecules to intracellular structures, cells, tissues, organs, organisms, populations, ecosystems and ultimately to the biosphere itself. Attitudes concerning the best way to explain important evolved features of this organizational hierarchy vary considerably, and take us to the heart of current disputes in theoretical biology. Certainly the question of group selection question is worthy of further consideration.

As Johnson notes, Dawkins is very dismissive of group selection. Dawkins rejection of group selection springs from his reductionist methodology (1987, p. 13):

> The hierarchical reductionist . . . explains a complex entity at any particular level in the hierarchy of organization, in terms of entities only one level down the hierarchy; entities which, themselves are likely to be complex enough to need further reducing to their own component parts; and so on . . . This was the point of explaining cars in terms of carburetors rather than quarks. But the hierarchical reductionist believes that carburetors are explained in terms of smaller units . . ., which are ultimately explained in terms of fundamental particles. Reductionism, in this sense, is just another name for an honest desire to understand how things work.

Dawkins thus has a tendency to cast explanations of biological phenomena at the genic level.

Selection, for Dawkins, is selection at the genic level, not the organismal level of the hierarchy, and certainly not the group level. The group phenomenon of altruism, for example, is explained as an indirect consequence of selection operating on selfish genes lower in the organizational hierarchy. As Johnson points out, this raises some interesting issues.

To explore this topic in more detail, let me begin by noting that properties called *adaptations* are the quintessential products of natural selection. Elliot Sober defines adaptation as follows (1993, p. 84):

> Characteristic c is an adaptation for doing task t in a population if and only if members of the population now have c because ancestrally there was selection for having c and c conferred a fitness advantage because it performed task t.

To use Sober's example, to say that the function of the heart is to pump blood, is to say that an organ was selected for which had this trait — it conferred some fitness advantage upon our (distant) ancestors. So adaptations can be differentiated from other properties of those biological systems, in that the former are the effects of natural selection.

The present puzzle is this: properties worthy of the name *adaptation* are found up and down the biological hierarchy. Since some groups seem to have properties classifiable as adaptations, are those adaptations a result of selection operating *at the group level*? Or is it a by-product of selection operating at, say, the genic level? These questions are hard to answer unless you clarify what is meant by *group* and *group selection*.

What exactly is meant by the term 'group selection'? According to Ernst Mayr, hard group selection (1997, p. 202):

> . . . occurs when the group as a whole has certain adaptive group characteristics that are not the simple sum of the fitness contributions of the individual members. The selective advantage of such a group is greater than the arithmetical mean of the selective values of the individual members. Such hard group selection occurs only when there is social facilitation among the members of the group or, in the case of the human species, the group has a culture which adds or detracts from the mean fitness value of the members of the cultural group.

Mayr's reference to a socio-cultural dimension of the group selection issue makes his remarks of relevance to one of Johnson's main examples.

Johnson, too, points to the socio-cultural dimension of group selection in his discussion of "male bellicosity," and the compensating response elicited from females. Rejecting a *sexual selection* model of explanation, Johnson's alternative hypothesis is that these traits arise from differential survival of groups — group selection will tend to weed out those groups in which bellicose males don't get preferential access to mates.

The main problem here is that Johnson simply rejects alternative explanations out of hand, and he provides no *evidence* for his preferred group selectionist hypothesis. (Unfortunately, in sociobiological circles, this sin not restricted to Johnson). It looks very much as though what we have here is a group selectionist "just so" story. This criticism applies, too, to his discussion of the phenomenon of *disapproving elders*. We have a hypothesis, but where is the evidence? A proper treatment of these examples would take us into the domain of anthropology and the study of cultural evolution. In the context of Johnson's actual example, the phenomenon of racism would also have to be addressed.

The phenomenon of aggression, and female responses to aggression, is in point of fact an extremely complex issue, both biologically and culturally. Johnson admits his scenario is simplified. But the devil lies in the details, and these Johnson does not discuss. I agree with Johnson that we should not be dismissive of group selectionist claims simply because they run contrary to prevalent ways of thinking in the biological sciences, but before such claims can be evaluated, they need to be worked out in much more detail, and then they must be tested.

Johnson's essay also contains the claim that groups are to be thought of as a new type of organism — perhaps *individual* would be a better word, given the structure of the organizational hierarchy. The claim is put forward and developed through a series of analogies. Again, the issue here is complex, with much literature being devoted to the question as to whether there are biological individuals above the organismal level of the hierarchy. My criticism here is that while questions surrounding group selection may have a bearing on the *individuals question*, they will underdetermine its resolution. For Johnson to deal effectively with this issue, he would have to say a lot more about what individuals are, what groups are, and why the latter are like the former in relevant respects.

In the end, I hope my disagreement is constructive. Johnson has opened a can of theoretical worms that probably needs to be opened, and often. Reductionism, while it has its place in biology, should not blind us to the complexity of biological phenomena, and the possible need to reach beyond the genic level in our explanations of such phenomena.

REFERENCES

Dawkins, R. (1987). *The Blind Watchmaker.* New York: Norton.
Mayr, E. (1997). *This is Biology.* Cambridge, MA: Harvard University Press.
Sober, E. (1993). *Philosophy of Biology.* Boulder, Colorado: Westview Press.

Niall Shanks
Department of Philosophy
Department of Biological Sciences
Department of Physics and Astronomy
East Tennessee State University
Johnson City, TN 37614
e-mail: shanksn@etsu.edu

Johnson's Reply to Shanks

Does the group selection take place? It seems to me that the existence of the groups (hives, armies, colonies, swarms) of the eusocial species provides strong evidence that it has taken place. My speculation is that accidental favorable co-adaptations within these groups have been selected for by survival differences *among the groups*, to the point that the surviving groups have now achieved organism status. The reductionist alternative explanation is that the social organization and differentiation within these groups can be explained entirely in terms of the reproductive interests of the individuals constituting the group (or the genes they bear). This type of explanation is profoundly fallacious in the case of the eusocials because of certain circumstances that mark the organismhood of these groups: they exhibit bottleneck reproduction, and differentiation of the different "tissues" of the group organism is brought about by treatments imposed by the organism upon the larval (or infant, or fetal) individuals of the group. Of course, the reproductive interests of the constituent individuals must have played a role in the evolution of these groups before they achieved organismhood. But it seems to me to be very unreasonable to claim that differential survival — selection — among groups played no role in that evolution, and that the burden of proof of such selection is on those that claim it exists.

Further on the question of the existence of selection at the group level, I contend that the existence of multicellular organisms, presumably arising from a world in which all organisms were single-celled, is evidence that something akin to group selection has taken place, and is, in fact, a spectacular force in evolution. The existence of eukaryotic cells, even the existence of any cells at all, provides similar evidence (assuming that the current accounts of how these things arose are correct).

How did multicellular organisms arise from a world of single-celled creatures? Are we obliged to eschew all mention of selection at the "group" level in discussing this miracle?

Regarding my 'bellicosity of young male humans' example, Professor Shanks is correct in noting that I have provided no evidence that the development of this trait was supported by selection at the group level. Nor could I. In fact, I do not know whether such selection played a role in the development of that trait, and I doubt that anyone will ever know for certain. My purpose was to present an instance in which the hypothesis of selection at the group level leads to a plausible account — a "just so" story. I see that I have failed to communicate my idea of the role of group selection, because Professor Shanks accuses me of rejecting "alternative explanations out of hand." I do *not* reject, for instance, the sexual selection explanation. Shanks here is committing precisely the error that I had hoped to do away with forever. To repeat the thesis buried in my essay, selection forces at various levels are not *opposed*, but *superposed*. They *add*, like force vectors. My speculation was that the preference of women for cavalier young men was good not only for them as individuals, but also for the "group" (the tribe, the nation), by providing young warriors with genetic compensation for their willingness to go to war. This was posited as a possible (but we will never know for sure) instance of accidental co-adaptation within a group that turns out to be advantageous to the group. Not that there will always be such a positive outcome in the adding of the forces of selection. For instance, if a voracious predator invades a territory inhabited by peacocks whose males have evolved to a state of helplessness because of the advantages in sexual selection of male display, it could well be the case that entire peacock population of the territory will be wiped out. This would be another instance of *addition*, with the disadvantage canceling the competitive advantages of heavy tails among the individual male peacocks. It will be objected that both the advantage and the disadvantage bear on the individual male peacocks. It will be objected that both the advantage and the disadvantage bear on the individual peacocks; yes, but their disadvantage is *also* a disadvantage to the group, the local peacock population, and could bring about the demise of that group.

I am baldly proposing the consideration of the superposition of selection forces at various levels because it is plain to me that such superposition takes place — groups do die or live, the individuals within them die or live if the groups live, and die or migrate if they don't. The doings of the individuals of the group do result in accidental co-adaptations increasing the fitness of the group, or in accidental "disease" conditions that decrease that fitness.

The force vector of group selection is small, and we must be wary of error in speaking of it. But that danger of error does not, in my opinion, justify the furious rejection of consideration of selection at the group level that one finds in Dawkins and Grafen, especially if one agrees that in the past, life forms have somehow assembled to make new compound life forms, and that this phenomenon constitutes evidence of *something* of interest going on that looks very much like group selection.

Poznań Studies in the Philosophy
of the Sciences and the Humanities
2000, *vol.* 71, *pp.* 168–181

Cory F. Juhl

Teleosemantics, Kripkenstein and Paradox

ABSTRACT. Recently Millikan has proposed a "teleosemantic" solution to
the "Kripkenstein problem." The latter problem is raised by Kripke on behalf
of Wittgenstein, and purports to show that there can be no satisfactory account
of how 'plus' means plus, or more generally how meaning is possible. Millikan
believes that evolutionary biology solves the problem. Her solution draws an
analogy between semantic norms and norms governing biological organs. A
human heart functions properly, according to evolutionary norms, when it behaves as it did in our ancestors in so far as those earlier behaviors aided in
their reproduction. Norms governing mental tokens, and even public-language
tokens, are supposed to be explicable along similar lines. In the paper I argue
that Millikan does not satisfactorily solve the problem of explaining how 'plus'
gets metaphysically hooked up to plus, as opposed to some other function. In
his reply Bonevac mentions an interesting paper by Miscevic which raises similar problems for Millikan, but attempts to shore up the Millikan solution by
appealing to a notion of "physically ideal conditions." In my reply to Bonevac
I briefly explain the Miscevic "teleocomputational" strategy, and why it fails.

1. Introduction

Recently a novel approach to the problem of giving a naturalistic account
of semantics has been developed. The approach, commonly referred to as
the "teleological" theory, attempts to account for semantic relations[1] in
terms of a notion of *proper function*. To a zeroth-order approximation,
the teleological view says that, just as our livers have proper functions
(where these functions are determined by our evolutionary history), our
brains or parts thereof have as their proper function to map onto states of
affairs in particular ways.[2] In this paper I will argue that the teleological
account does not, contrary to what Millikan asserts, solve the Kripkenstein
paradox.[3]

---

[1]For the purposes of this discussion, I will not distinguish between semantic and
representational norms.

[2]The most detailed account of Millikan's view is given in her (1984). A useful
summary is in Millikan (1989a). Millikan (1989b) contains an extended discussion of
the notion of a proper function. David Papineau (1993) has developed a similar theory.

[3]In my opinion the best overview of matters related to the Kripkenstein paradox is
Boghossian (1996). Kripke (1982) is the ur-text.

Exactly what the "Kripkenstein paradox" is supposed to be is controversial, and the issues raised by Kripke (1982) have been discussed at length elsewhere. Rather than engage in this controversy here, I will just state what I take to be the upshot of the "paradox." The paradox is that there appears to be no way to explain how semantic norms are possible, since no facts, whether facts about individual or community dispositions or some other kind of fact, could possibly make it the case that a given application of a term (whether of a public language or "mentalese") is correct or is incorrect. Consider the word 'plus'. Some statements, e.g., 'Two plus two equals four' are true, or correct, statements (in English), whereas 'Two plus two equals five' is not. One problem facing the would-be naturalist semanticist, is to make sense of the notion of 'ought', the normative features, associated with uses of 'plus'. I will refer to this, following Boghossian (1989), as the "normative problem." Another aspect of the problem is the "descriptive problem": any correct account of semantic relations, naturalistic or otherwise, must get the extension of terms right. For example, if an attempted solution entails that $3 + 3 = 6$ or the sum of 3 and 3 is indeterminate fails descriptively, whatever its merits with respect to the normative problem. The problems raised in this paper primarily concern the descriptive problem, although the two problems are interestingly related.

The basic difficulty that I raise for Millikan's theory is that our finite evolutionary history together with finite learning histories of individuals and/or their communities is insufficient to connect a given representing system with a unique infinitary object like the addition function, whether we are considering public tokens or "internal" ones within brains. This point will be made more precise, but before proceeding let's look at an example of Millikan's, the hoverfly and its "rule."

The male hoverfly apparently spends much of its time hovering around (hence the name) until it detects something moving within its field of vision. When such a moving object is detected, the hoverfly is genetically programmed to fly in a direction and rate determined by the velocity of motion of the retinal image. Leaving aside details of the story, Millikan claims that the proper function of a certain subsystem of the hoverfly is to conform to a rule of the form 'Fly in a direction X when Y occurs'.

For simplicity, I will disregard Y, by considering a simpler case in which flies (perhaps those from some other even dumber species) "ought to" always veer off at an angle X no matter what. What is the precision of the angle X? Some qualitative considerations, in particular the fact that only finitely many past events would seem to be relevant to the account, and that only finitely many relations to finitely many other objects would appear to be relevant as well, suggest that only finitely many angles are definable from what can be legitimately appealed to (unless infinitary, e.g., "recursive" constructions are allowed, a possibility to be discussed below). But the particular finite set will depend upon exactly the historical sequence of events, in general. So why should we expect a precise angle X, from a prima facie uncountable (at least) collection of candidates, to be among those definable in terms of the historical facts?[4]

The claim that only finitely many angles are definable from finitely many past facts requires that the teleologist does not have at her disposal limit constructions, or in general notions of definability that allow infinitely many different quantities to be defined in terms of finitely many. Limit constructions are infinitary objects, so we must assume that some recursion rule, or whatever infinitary object is used to determine the limit construction, is uniquely picked out by finitely many historical events. But this is just another special case of the problem that we are worried about. If we are to satisfactorily explain how a finite history selects an infinitary item from an infinite class, then we may not presuppose in our account that the problem has been solved.

However, Millikan appears to be presupposing just that. In her (1990), she considers a related objection, as follows.

> To say that the hoverfly has as a biological purpose to follow the proximal hoverfly rule is also quite different from saying that this rule is the only rule that fits all past instances of hoverfly turns, say, that resulted in hoverfly pro-creation. Suppose it were so that never in history had a male hoverfly spotted a female that happened to approach him at such an angle as to produce an. . . angular velocity between 500 and 510 degrees per second. Then the proximal quoverfly rule, [a "Goodmanized" version of the proximal rule that Millikan endorses], fits all past actual cases of successful female encounters. But it is not a rule the hoverfly has as a biological purpose to follow. For it is not because their behavior coincided with that rule that the hoverfly's ancestors managed to catch females, and hence proliferate. In saying that, I don't have any particular theory of the nature of explanation up my sleeve. But surely, on any reasonable account, . . . [the unnecessary complexity in the Goodmanized rule] is not a functioning part of the explanation. . . (True, I am making the assumption [that the Goodmanized rule is genuinely more complex]. . . But my problem is to solve the Kripke-Wittgenstein paradox, not to defend common-sense ontology. Nor should either of these projects be confused with solving Goodman's paradox.) (1990, p. 333)

Her response to Goodmanized rules suggests a similar response to the worries I have enunciated. Perhaps Millikan would dismiss all but one of the infinity of alternatives as essentially "Goodmanized rules." What rules them out as possible alternative proper functions is that they are not involved in *genuine explanation*, whatever that turns out to be.

---

[4]Letting X be an interval does not help, since the endpoints must then be precise. More complex constructions do not help so long as there are infinitely many initial possible "rules," as will be shown in problem 2.

It will be important to distinguish between syntactical objects, what we might call rule-specifications, from the rules themselves. I leave aside the question of what rules are on a Millikanian view, but note that Millikan must insist that rule simplicity closely mirrors rule-specification simplicity. Otherwise there would be no reason to infer from the (alleged) fact that the Goodmanized rule-specification is simpler than the non-Goodmanized rule-specification the claim that the corresponding rules are similarly ordered with respect to simplicity. (It may be useful for heuristic purposes to think of a rule as a collection of behaviors, or perhaps a "correctness relation" between conditions and behaviors, whereas a rule specification is a linguistic token or type.)

2. Two Problems

Problem 1. *Without an account of simplicity and explanation, Millikan does not solve the rule-underdetermination problem.*

The point of the Millikanian enterprise is to give a naturalistic account of semantic norms. On Millikan's view, explanatory relations are governed in part in simplicity considerations. If simplicity and explanation are going to be invoked to ontologically determine which of a presumably infinite variety of possible norms actually applies to a given biological system, then these relations must be objective, "natural" relations. So it would seem that within the natural world we have an objective simplicity ordering of rules, and it turns out that the simplest rule that is explanatorily compatible with (i.e., has the right sorts of explanatory relations to) past history is the proper function or rule of behavior. If this is the account, then it is difficult to see why the simplicity ordering invoked is any better than a solution to the problem by fiat. Without a theory of simplicity, and a plausible account of why we ought to think of such simplicity relations are present within the natural world, they seem to be little more than an ad hoc "magical" device for solving an otherwise intractable problem. It is analogous to a "solution" to Goodman's problem by simply saying that "green" and "blue" are simpler than "grue" and "bleen," and leaving things at that.

Alternatively, one might define simplicity in terms of a more fundamental explanation relation, so that the simplest rule is just defined to be whatever genuinely explains the relevant phenomena. However, this move is problematic as well for Millikan, given that she tries to account for the distinction between explanations in terms of simplicity. Without an independently characterized notion of simplicity, there remains of Millikan's argument against Goodmanized rules only the bald claim that the one rule genuinely explains and the other doesn't. So this alternative is equally difficult to distinguish from solution by ad hoc fiat. In short, without a plausible account of simplicity (a notoriously elusive entity) and/or explanation, the problem of specifying what makes one infinitary object rather than another the representandum of a biological system remains untouched.

Not only does the problem remain unsolved, it remains difficult to see how it could possibly be solved by any account of the sort proposed by Millikan, for the following reasons.

Problem 2. *The Teleological account faces a Historical Heap Problem.*

At some stage of the history of the universe, presumably, nothing had addition as its proper function. But at some stage of evolutionary history, some system gets addition as its proper function. So consider the following argument:

P1: One new historical event is insufficient to pick out addition as the proper function associated with 'plus' from infinitely many alternatives, if addition is not already determined as one among at most finitely many candidates for the role of the proper function before the event.

P2: Addition is one of infinitely many candidate possible proper functions present at some initial stage.

P3: Only finitely many relevant historical events have occurred by any stage of history.

Then by induction, addition is never selected as the unique proper function of 'plus' (as before, either a mental token/type or a public token/type).

Which of the assumptions P1–3 may be plausibly rejected? A Millikanian might respond that there are only finitely many candidate proper functions initially, i.e., deny P2. Further, plus (fortunately!) happens to be among this finite set. It is difficult to consider this proposal in detail without a precise account of the ontology we are supposed to have at our disposal. However, it is hard to imagine a plausible view along these lines, since the reason that 'plus' might seem likely as a candidate is that it is recursively definable from some other plausible candidates. But if we allow that addition is a candidate because of its recursive specifiability, then how are we to motivate ruling out the infinitely many other recursively specifiable rules? One attempt might appeal to the fact that for some recursive specifications RS, it might take a hard disc larger than the known universe to store the code for RS, and for any such rule, or any rule requiring more complex (in terms of Kolmogorov-Chaitin complexity,[5] say) specification, may be ruled out as a candidate.

This reply conflates rule-specification complexity and rule complexity. For we have been given no reason to think that a rule that cannot be specified finitely at all should not be the rule governing the behavior of some system, the proper function of that system. If explanatory relations are simply a part of nature, why shouldn't an arbitrarily complex (in the sense that all of its rule-specifications are sufficiently complex) rule be associated with a system? That is, why should we think that every explanation of the relevant sort is of finite complexity (can be finitely expressed in some language)? Once we move explanatory relations into the realm of nature, existing (in the majority of cases) independently of any human activity, the usual appeals to pragmatic concerns seem irrelevant to answering such questions.[6]

Alternatively, a Millikanian might suggest that the argument raises nothing more than the paradox of the heap, a problem for all vague predicates. Representation is a vague relation, it might be suggested, and so we may pick some standard solution to the paradox of the heap (whichever one best fits the pertinent naturalistic project). The main problem with this response is that it basically concedes the game to the opponent who was worried about why exactly one function out of infinitely many is uniquely determined by the historical facts. For the response cannot be the claim that 'plus' is vaguely or partially connected to infinitely many functions at all stages in history, because that is precisely the consequence that the response must be designed to avoid (unless a sort of finitist position, described below, is adopted). If the response is that at some stage in history only finitely many functions are vaguely related to 'plus', or that after some single historical event all vagueness is removed when infinitely many functions were vaguely connected to 'plus' before the event, then essentially the same problems arise. Variants of the above "heap" argument can be constructed in obvious ways, leading to essentially the same concerns.

One might deny P3, and claim that infinitely many relevant events have occurred. I fail to see how to make a go of this response, since

---

[5] The Kolmogorov-Chaitin complexity of a sequence of objects is, roughly, the length of the shortest program which generates exactly that sequence.

[6] It may seem to some that I am attributing an absurd position to Millikan when I say that for her, explanatory relations exist in nature independently of human activity. However, explanatory relations are what do the work for her in determining representation relations between, say, hoverfly parts and their environments. These were in place before any human was in existence. Even if we say that the relations involved are really more like causal relations that play crucial roles in explanations, the arguments that I have given are unaffected.

the standard examples of relevant happenings are events like a hoverfly's flying off in a given direction when retinally stimulated in a particular way. So perhaps a better response is to deny P1, that a single historical event cannot pare down an infinite set of candidates to a finite set. I do not know of a way to refute this contention. But again, given the sorts of examples standardly given, I do not know of a plausible candidate for such an event. It simply won't do to claim that when a large number of flies perform a large number of hoverings, magically the problem is solved, even though no single hovering yields the required infinite collapse of the set of possibilities.[7]

A response to the historical heap problem that a semantic teleologist might make is that, indeed, 'plus' doesn't mean something infinitary after all. In fact, the addition rule only governs some finite set of possible applications (or application attempts). 'plus' is indeterminate in meaning, in the sense that it has no bearing on, say, pairs of numerals whose decimal expansions cannot be written down on a tablet the size of the physical universe. Rules that can govern any physical entity's behavior can have no bearing on physically impossible behaviors. This sort of radically revisionary semantic theory leads to a hornet's nest of difficulties that are impossible to address in a paper this short. I will simply mention that if this response is adopted, then the view inherits many of the problems of extreme finitism as a philosophy of mathematics. We might ask, for instance, how plausible it is to believe that there is some largest natural number that has a unique successor. 1 has a unique successor, perhaps Millikan will grant. But since eventually, for some n, the successor function, like plus, must fail to have a determinate application to n, there must be a particular last number with a determinate successor. Mathematics and logic will have to be radically revised, since some axioms of arithmetic are false on this account, as are axioms in the metatheory of first-order logic (since inference rules, for example, are of the same infinitary nature as 'plus'). These difficulties and others that arise from such radical revisionism make this sort of response highly unattractive, to be avoided except possibly as an absolute last resort. I might add that it would be rather ironic if Millikan, who attacks Dummettian "verificationism"[8] as a misguided consequence of "meaning rationalism" (her arch-enemy), were forced into a view normally associated with the most extreme form of such verificationism, strict finitism.

The argument can be recast somewhat, so that the focus is on the infinitary precision of the plus-rule. That is, 'plus' seems to carry with

---

[7] Godfrey Smith (1994) suggests that it is unclear whether explanatory relations are sufficient to determine a unique representation, but does not pursue the matter further.

[8] See, e.g., Millikan (1990).

it norms governing infinitely many possible clearly distinguished circumstances (for example, the different possible number-pair "presentations") and responses appropriate in those circumstances. The analogous question becomes, how can a finite sequence of relevant historical events yield a norm of infinitary precision, in the sense specified? Although such an infinitary norm may be vague, in that there may be a continuum of cases in between different circumstances or correct responses, such vagueness is compatible with infinitary precision in the sense just described. The fact that this sort of difficulty arises and appears insoluble by the teleosemantic approach is, I believe, just one symptom of the fundamental misguidedness of any attempt to account for semantic norms as a byproduct of causal history. This is a matter, however, for another essay.

REFERENCES

Boghossian, P. (1989). The Rule-Following Considerations. *Mind* **98**, 507–549.

Godfrey-Smith, P. (1994). A Continuum of Semantic Optimism. In S. Stich and T. Warfield (Eds.), *Mental Representation: A Reader.* Oxford: Blackwell. pp. 259–277.

Kripke, S. (1982). *Wittgenstein on Rules and Private Language.* Cambridge, MA: Harvard University Press.

Millikan, R. (1984). *Language, Thought, and Other Biological Categories.* Cambridge, MA: MIT.

Millikan, R. (1989a). Biosemantics. *Journal of Philosophy* **86**(6), 281–302.

Millikan, R. (1989b). In Defense of Proper Functions. *Philosophy of Science* **56**, 288–302.

Millikan, R. (1990). Truth Rules, Hoverflies and the Kripke-Wittgenstein Paradox. *Philosophical Review* **94**, 323–53.

Papineau, D. (1993). *Philosophical Naturalism.* Oxford: Blackwell.

Cory F. Juhl
Department of Philosophy
University of Texas at Austin
Austin, Texas 78712-1180
e-mail: cjuhl@mail.la.utexas.edu

Commentary by Daniel Bonevac

Cory Juhl underestimates his own "historical heap" argument. A finite number of grains of sand makes a heap, despite a sorites argument. But finitely many foreclosures of possibilities each of which allows infinitely many to remain never narrows the possibilities to a finite collection. Juhl's argument may be recast as a straightforward weak induction. Assume

that episodes that affect the interpretation of '+' are ordered historically
in stages.

Assumption. Nothing other than the history of the use or users of '+'
can foreclose possible interpretations of '+'.

Theorem. At every finite stage of history, there remain infinitely many
possible interpretations of '+'.

Proof. By weak induction on stages of history:
Basis: At the first stage, there are infinitely many possible interpretations
of '+' consistent with the history of its use or users, since there is no such
history, and nothing else can foreclose possible interpretations.
Induction: Assume that there are infinitely many possible interpretations
of '+' at stage n. The episode leading from n to $n + 1$, however, cannot
foreclose all but finitely many possible interpretations of '+'. And nothing else forecloses possible interpretations, by assumption. So, there are
infinitely many possible interpretations of '+' at stage $n + 1$. Say that a
historical theory of meaning adopts the above assumption, and that the
meaning of '+' is radically indeterminate if it has infinitely many possible
interpretations at any historical stage. Then Juhl's argument shows that
every historical theory of meaning makes '+' radically indeterminate in
meaning.

As Juhl argues, however, maintaining that the meaning of '+' is radically indeterminate is hard to square with mathematical theory or practice. Not only would the meaning of '+' be indeterminate now, despite
the decidability of sentences of arithmetic that lack unbounded quantifiers
and the categoricity of second-order arithmetic; nothing we could say or
do could remove the indeterminacy. If this seems absurd — if, in short,
we are not to surrender to Kripke's sceptic — then Juhl has given us a
knock-down argument against historical theories of meaning. The advantage of casting the argument in this form is that it shows that the only
point at which the historical theorist can demur is at the inductive step.
(The basis is justified by the assumption and the existence of infinitely
many logically possible interpretations. And there should no longer be
any temptation to search for a sorites fallacy.)
    Millikan evidently holds that some historical episodes can foreclose all
but finitely many possible interpretations. As Juhl points out, however,
her appeals to simplicity and explanatory power are at best programmatic
and at worst question-begging. We need to know why interpreting '+' as

plus is explanatory in a way that interpreting it as quus is not explanatory of a history fully consistent with both interpretations. This form
of the argument also shows why certain common moves cannot succeed.
Appeal to community practices does not help; the history of our use of
the term fares no better the history of my use of the term, unless there are
infinitely many of us (compare Blackburn 1984). Appeal to evolutionary
history as fixing proper functions cannot help, for the argument does not
distinguish historical episodes of the use of the term from other kinds of
historical episodes. Moreover, not only are historical theories inadequate
in themselves, on pain of radical indeterminacy, but history can play only
a supporting puzzle-solving (i.e., determinacy-producing) role in a theory
of meaning for terms such as '+'. Appeal to history cannot rule out all
but finitely many possible interpretations of '+'. Other factors must be
responsible for singling out one from infinitely many candidates. In any
mixed, partially historical theory, therefore, the significance of history is
no greater than the significance of those other factors. Indeed, unless the
historical and nonhistorical factors mesh in just the right way, history will
play.

This suggests that Millikan's theory cannot easily be supplemented
with extra-evolutionary considerations to solve the Kripkenstein puzzle.
Nenad Miscevic, for example, supplements Millikan's teleological theory
with physicalistic constraints. His teleo-computational view relies on the
point "that mathematical operations are performed by means of physical
operations" (1996/97, p. 182) and thus that, if purely physical dispositions
are determinate, mathematical dispositions are too. Imagine a calculator
that adds (or quadds):

> there is a discoverable matter of fact whether the calculator has a disposition
> to perform the operation for any arbitrary number of steps. This matter of fact
> is determined by the precise calculational makeup of the device, and it can be
> ascertained at least in principle by standard methods, including induction and
> unproblematic idealizations (1996/97, p. 185).

Now I doubt that Kripke would grant that purely physical dispositions and
standard methods such as induction and idealizations concerning memory
size, etc., are unproblematic from the sceptic's point of view; the sceptic might turn his attack to calculators as well as humans (Hsu 1993).
At best, it seems, noting that mathematical operations are performed by
physical operations might reduce the Kripkenstein puzzle to the Goodman puzzle, at least for certain simple mathematical expressions. But the
point on which I wish to focus here is that, to the extent that Miscevic's
theory succeeds, it succeeds because of its physicalistic and computational features, not because of its teleological features. In short, in the

teleo-computational approach, the computational part does all the puzzle-solving work; evolutionary history plays no role at all. This is just what Juhl's argument should lead us to expect.

Causal theories of meaning and reference have dominated much work in semantics and the philosophy of language for the past three decades. Juhl's argument raises an important obstacle to extending such theories beyond the realm of proper names and natural kinds. Kripke's (1970) or Evans' (1973) causal theories of names survive Juhl's argument, for an act of baptism can single out one interpretation. The same is true for natural kind terms, given fairly weak constraints on the underlying kinds. But Juhl gives us reason to beware attempts to extend causal theories to terms such as '+', 'causation' (Horwich 1985), or other abstract terms. Epistemic access to addition or causation is more problematic than epistemic access to Nixon or gold. Moreover, no accompanying sortals provide help. Critics have worried that ostensions do not themselves suffice to pick out a determinate referent; baptism and rigid designation rely on an accompanying sortal, as in Geach (1967). To baptize an entity as 'Nixon', one must specify whether 'Nixon' refers to that man, or that mass of molecules, or that shape, etc. Whether that is true or not, stipulating that Nixon is a man, for example, forecloses many otherwise possible interpretations. The same is true of gold and chemical substance. But nothing plays the same role for addition or causation. Addition is a function on numbers; causation is a relation among events and states. But there are infinitely many relations of these kinds, and nothing akin to ostension to single out one of them.

REFERENCES

Blackburn, S. (1984). The Individual Strikes Back. *Synthese* **58**, 281–302.
Evans, G. (1973). The Causal Theory of Names. *Aristotelian Society Supplement*, **47** 187–208.
Geach, P. (1967). *Reference and Generality*. Ithaca, NY: Cornell University Press.
Horwich, P. (1985). *Causation*. Cambridge, MA: MIT Press.
Hsu, A. (1994). *On Some Remarks of G. E. M. Anscombe's Concerning Kinds*. Ph.D. dissertation, UCLA.
Kripke, S. (1970). *Naming and Necessity*. Cambridge, MA: Harvard University Press.
Miscevic, N. (1996/97). The Skeptic and the Hoverfly. *Acta Analytica* **16**, 171–187.

Daniel Bonevac
Department of Philosophy
University of Texas at Austin
Austin, Texas 78712-1180
e-mail: dbonevac@mail.la.utexas.edu

Juhl's Reply to Bonevac

Professor Bonevac and I are largely in agreement. Since the paper that he is responding to was written, he and I learned of the existence of a fine paper by Nenad Miscevic (see reference in Bonevac's reply) that raises many of the same objections to Millikan's approach. Miscevic believes that he has found a genuine solution to the Kripkenstein paradox, a solution which supplements Millikan's attempt in an interesting way. However, Bonevac's claims, "to the extent that Miscevic's theory succeeds, it succeeds because of its physicalistic and computational features, not because of its teleological features" (p. 177). Although this may sound odd, given Miscevic's own belief that teleology is a crucial component of his solution, Bonevac is right to emphasize that Miscevic's proposal, if successful at all, solves only the descriptive problem, to which teleology is essentially irrelevant even on Miscevic's view. In what follows, I will briefly outline Miscevic's strategy and then show why the strategy seems hopeless.

Miscevic's solution, in a nutshell, is as follows. Suppose that there is a physical system that is disposed to add for any possible argument, i.e., such that for any pair of numerals input, the numeral for their sum is output. While Miscevic grants that any real machine breaks down eventually, we can in principle specify physically ideal conditions under which the machine is disposed to add absolutely any pair of numbers. The switch to *physically* ideal conditions is crucial, as most appeals to "ideal conditions" seem to presuppose the norm that is supposed to be constructed, and hence violate the constraints of the enterprise. But if we can specify physically ideal circumstances in a way that doesn't appeal to addition, as Miscevic claims, then the dispositionalist has overcome one of Kripke's main obstacles.

Once we have granted the physical possibility of a system that is disposed to add, we are to suppose that one of our biological organs is such a system. If the "addition organ" was selected for, then its proper function is to add. Proper function, acquired via the right causal history, is what is supposed to introduce *norms* into the picture.

My objection to Miscevic's proposal is that even if he solves the descriptive problem, he does so in a way that guarantees failure on the normative problem. I think that he does solve the descriptive problem, in a narrow sense. For consider a physical instantiation of a Turing Machine that is supposed to add. For definiteness, suppose that the TM has two input tapes on which it reads standard binary numerals right-to-left, and an output tape on which it writes binary numerals right-to-left. The read/write head is a finite-state device. Let us define 'physically ideal conditions' as physical conditions in which a particular set of state transitions occur, where the states are all specified physicalistically. Let the state transitions be exactly those that lead to an outputting of the sum-numeral of two input numerals. No appeal has been made to addition. Hence, it seems that in this way we may specify, in a licit way, 'physically ideal conditions' such that, when they hold, the machine adds. So there seems after all to be a sense in which the machine is "(physically) ideally disposed" to add.

If this is how we interpret this part of Miscevic's proposal (although it is unclear whether Miscevic has this in mind, given certain later statements and his belief that it gives aid to the teleosemantic cause), then we seem to have overcome a thorny difficulty that has beset quite a few dispositionalist approaches. There is a sense in which we can say that certain physical machines are ideally disposed to add. But it is important to note important differences between such "ideal" dispositions and dispositions characterized in terms of what systems will (probably) actually do. First, supposing the correctness of quantum mechanics, it seems probable that any real agglomeration of matter/energy that is designed for adding will eventually screw up. For long enough numerals, state-transition probabilities for chunks of quantum matter are such that any physically possible machine will, with high probability, eventually make a transition that is not among those we have included in our list of "physically ideal conditions." That is, it seems that for any physical system, there exist numerals such that the system will with probability greater than $1/2$ fail to add them correctly. In this sense no physical machine is disposed to add. This is the sort of consideration that has been taken to undercut dispositionalist attempts to solve the descriptive problem. However, the fact that no real device is disposed (in this latter sense) is irrelevant to the Miscevic proposal under consideration. A physical machine can be Miscevic-disposed to add, yet probability-disposed to screw up. What Miscevic shows us is one way to fill-in the "ceteris paribus" clause that dispositionalists had been unable to specify previously.

It might seem that the teleological part of Miscevic's solution proceeds unproblematically as well. For if we have a physical device in our brain that is something like a TM, and adds for small arguments, then it is ideally disposed to add in some well-defined sense. If, furthermore, the state-transitions required for small-argument sums were selected for (the environment is such that systems that add wrong on small arguments are eliminated), to undergo such state-transitions is the proper function of the system. However, if we look more closely, the proposal loses its luster.

One way to make the main problem apparent is to note that virtually any physical object is "ideally disposed" to add. That is, we can specify "physically ideal conditions" under which the cup on my table would add (specifying what counts as input and output of numerals). Unfortunately, for the same reason virtually every object is "ideally disposed" to quadd as well. Even if we want to quibble about some objects, it seems clear that any middle-sized object used as a Turing machine will have a large, possibly infinite, number of "ideal dispositions." We get different notions of "ideally disposed" for different specifications (stipulations) of physically ideal conditions. It should be clear that there can be no selection pressure for or against any "ideal disposition" of this sort. So such Miscevic-dispositions are actually no help to the teleosemanticist in accounting for how an organ might acquire addition as opposed to quaddition as a norm governing its behavior.

These considerations help us to see that the teleosemanticist faces horns of a dilemma if she wants to avoid finitism. Either our brains must be probability-disposed to add (which seems impossible, given QM), or they are merely ideally disposed, for some sense of "ideal conditions." (Biologically Normal conditions play this role in Millikan's account.) But it seems clear that there are many different ideal dispositions, one for each notion of biologically Normal conditions. This returns us to the central problem raised in the paper: How are just the right conditions picked out by the metaphysical hand of Nature to enable us to add rather than quadd?

Acknowledgments. Special thanks to Dan Bonevac for writing a stimulating response, and also to Nicholas Asher for helpful discussions of the Miscevic paper.

Poznań Studies in the Philosophy
of the Sciences and the Humanities
2000, vol. 71, pp. 182–218

Daniel Bonevac

CONSTITUTIVE AND EPISTEMIC PRINCIPLES [1]

ABSTRACT. Aristotle's account in Metaphysics Δ implies that there are
three kinds of essential properties: those that inhere in their subjects (a) of
necessity, (b) because of what their subjects are, and (c) usually. The first
kind is well-represented by the strict conditional. The second corresponds
to constitutive principles; the third to epistemic principles. Such principles,
though commonplace, have no representations in standard logical systems.
This paper constructs a logical theory of constitutive and epistemic principles
that illumines their relations to each other and to principles expressing accidental and essential predication. Constitutive principles assert that satisfaction of
the subject term is responsible for satisfaction of the predicate term. Epistemic
principles assert that satisfaction of the subject term is a good indicator of satisfaction of the predicate. I analyze constitutive principles as lawlike epistemic
principles. Constitutive principles imply epistemic ones, but only defeasibly.

We organize our thought with general principles:

(1) a. Acids are corrosive.
  b. Handguns are dangerous.
  c. Promises ought to be kept.
  d. Contracts must be honored.
  e. What goes up must come down.
  f. Patients with fungal respiratory infections have
    serious underlying illnesses.
  g. Accident victims who are cool to the touch may be in
    shock.
  h. Bulbs that do not light are burned out and should be
    replaced.

Such principles are commonplace. Oddly, however, they have no representations in standard logical systems. They are not extensional; they
are more than just accidental generalizations about what happens to be
the case, for they purport to give guidance about circumstances not yet
encountered. But they are neither universal nor necessary, for they admit

---

[1] I am grateful to Nicholas Asher, Michael Morreau, Marvin Belzer, Mark Lance,
Barry Loewer, and Paul McNamara for many helpful discussions over the past several
years.

exceptions. Space probes go up but do not come down, toy handguns are not dangerous, some contracts are illegal, etc. Inferences from principles to particular conclusions, therefore, are not deductively valid a fact which has led some to conclude that principles, despite their ubiquity, are of little use (Dancy 1991).

Moreover, they fall into at least two kinds. Some assert that satisfaction of the subject term is responsible for satisfaction of the predicate term. Something's being a contract, for example, makes it obligatory to honor it. I shall call these principles *constitutive*. Others assert that satisfaction of the subject term is a good indicator of satisfaction of the predicate. Fungal respiratory infections are good indicators of underlying illness, for example, but are not responsible for it. I shall call these principles *epistemic*. My goal in this paper is to construct a logical theory of constitutive and epistemic principles that illumines their relations to each other and to principles expressing accidental and essential predication.

1. Essence and Accident

Aristotle generally treats essential predication as involving necessity. "Essential attributes must inhere in their subjects of necessity," he says (*Posterior Analytics* 73b24); "it is impossible for them not to inhere in their subjects" (73b18). The sort of necessity involved, moreover, seems to be analytic or metaphysical. Essential properties are those that would have to be mentioned in defining a thing or kind (73a34-38; *Topics* 101b39; *Metaphysics* 1029b19-21, 1030a6-7, 1031a12-14, 1042a18). No surprise, then, that contemporary discussions of Aristotelian essentialism proceed in terms of analytic or metaphysical necessity.

Intriguingly, however, Aristotle at times distinguishes essential and accidental properties differently. First, he sometimes suggests that the kind of necessity involved may be causal rather than analytic or metaphysical: ". . . a thing consequentially connected with anything is essential" (*Posterior Analytics* 73b10). His example is surprising: "if a beast dies when its throat is being cut, then its death is also essentially connected with the cutting, because the cutting was the cause of death. . ." (73b13-15). Second, he sometimes suggests that necessity is not required at all. Consider his definition of "accident" in Book Δ of the *Metaphysics*:

> "Accident" means (1) that which attaches to something and can be truly asserted, but neither of necessity *nor usually*, e.g., if some one in digging a hole for a plant has found treasure. This the finding of treasure is for the man who dug the hole an accident; for neither does the one come of necessity from the

other or after the other, nor, if man plants, does he *usually* find treasure. And a musical man *might* be pale; but since this does not happen of necessity *nor usually*, we call it an accident. (1025a13-20; emphasis added)

If "essential" and "accidental" are contradictories — and Aristotle implies that they are (*Posterior Analytics* 73b3, 75a18-22) — then a property counts as essential if it inheres in its subject usually, even if not necessarily.[2]

This implies that there are at least three kinds of essential properties, in Aristotle's *Metaphysics* Δ view: (a) those inhering in their subjects with analytic or metaphysical necessity; (b) those inhering in their subjects with causal necessity — better, inhering in their subjects "according to nature," in Aristotle's phrase from *Physics* 192b35-39 — but as causal or physical consequences, not analytic or metaphysical ones; and (c) those inhering in their subjects usually, but not with any sort of necessity. The contrast between these kinds of essential properties is found in each of the following sets of sentences:

(2) a. Humans are animals.
     b. Humans are rational.
     c. Humans are omnivorous.

(3) a. Chairs are seats.
     b. Chairs have backs.
     c. Chairs have four legs.

(4) a. Promises are commitments.
     b. Promises ought to be kept.
     c. Promises are easily forgotten.

These are all *indefinite*, to use Aristotle's term (*Prior Analytics* 24a17, 19–21), or *generic*, to use a modern one. They are not universal; they are compatible with the existence of counterexamples. A single unforgettable promise, for example, does not falsify (4)c. A lone irrational human does not falsify (2)b.

In each case, the first sentence is in some sense necessary. The second is not; there are irrational humans, beanbag chairs, and overridden promises. But humans are rational because they are human. Chairs, by virtue of being chairs, have backs. And promises, *qua* promises, ought to be kept.

---

[2]The *Metaphysics* Δ account thus differs significantly from those found in *Metaphysics Z* — generally considered Aristotle's mature view — or *Posteriot Analytics I*, 4. In the terminology I introduce below, the *Posterior Analytics* account treats essential properties in the widest sense; the Z account, in the narrow sense; and the Δ account, in the widest sense.

The third sentence in each case expresses a true generalization, but one without any kind of necessity or causal force.

The same contrast applies to conditionals and generalized conditionals:

(5)   a.   If that is an insect, it is an animal.
         b.   If that is an insect, it has six legs.
         c.   If that is an insect, it has wings.

(6)   a.   If I salute Bob, I acknowledge Bob.
         b.   If I salute Bob, I show Bob respect.
         c.   If I salute Bob, Bob salutes me.

(7)   a.   If x cuts y's throat, x injures y.
         b.   If x cuts y's throat, x kills y.
         c.   If x cuts y's throat, x intends to kill y.

On one reading of these conditionals — at least (5)–(7)b and c — they too are indefinite in the sense that they are compatible with the existence of counterexamples. A lone beetle with a missing leg does not, on that reading, falsify (5)b; a single accidental throat-slicing does not falsify (7)c.

This has an important consequence for the logic of such conditionals: modus ponens does not hold. Given that Ralph is an insect and (5)b, it is reasonable to conclude that Ralph has six legs, but it is not guaranteed. Ralph may be unusual. Michael Morreau (1997) calls such conditionals *fainthearted*. Implicitly, they are qualified by a proviso such as 'provided conditions are suitable', 'other things being equal', 'in the absence of other factors', or the like.

Again, the first sentence in each case is necessary. The second is not; insects can lose legs, and there are sarcastic salutes and successful tracheotomies. Nevertheless, there is in each something like a causal connection. The third sentence in each case is true, generally speaking, but without any necessity or causal link.

Philosophers from medieval to modern times have tended to analyze essential predication as necessary and accidental predication as contingent. They have, in other words, treated all essential predication as being akin to (2)a–(7)a. Sentences such as (2)b–(7)b and (2)c–(7)c have received relatively little attention. And the contrast between them has scarcely been noticed.

My task here is to develop a theory of principles — that is, generics and conditionals — such as (2)b–(7)b and (2)c–(7)c. The former are *constitutive*, the latter, *epistemic*. In constitutive principles such as 'If p, then q', the truth of p does not metaphysically necessitate, but nevertheless is responsible for, the truth of q. If p is true, then q is true *because*, or *by virtue of the fact that,* q is true. In epistemic principles, this is not so. The truth of p provides evidence for the truth of q, if such a principle is true, but it is not itself responsible for the truth of q. I claim no originality for the theory of epistemic principles I adopt; it is Nicholas Asher and Michael Morreau's system of commonsense entailment. (See Asher and Morreau 1991, 1995; Asher 1995; Morreau 1997.) My contribution is the distinction between these kinds of principles and the analysis of constitutive principles in terms of epistemic ones.

2. Moral Principles

Asher and I (1996, 1997) have argued that, in moral reasoning, there is an important difference between epistemic and constitutive principles. The former are generalizations about obligation, permission, and other aspects of moral valuation. The latter are usually expressed by *ceteris paribus* principles. Promises, all other things being equal, ought to be kept. The principle, as usually understood, is constitutive; promises ought to be kept because they are promises. If a cynic maintains that promises ought to be ignored, however, the principle being advanced is probably epistemic. Promises ought to be ignored because they are easily forgotten, for example, or often insincere, not because they are promises.

Examining the contrast between constitutive and epistemic principles in moral contexts helps to clarify the logical relation between them. Epistemic principles, plainly, do not entail constitutive principles; that p's truth is generally good evidence for q's truth does not entail that p's truth is *responsible* for q's truth. The reverse inference, however, has much more plausibility. If the truth of p is generally responsible for, or brings about, the truth of q, then, ordinarily, the truth of p is good evidence for the truth of q. As I shall argue with respect to the examples to follow, inferences from constitutive to epistemic principles are not deductively valid. It is possible to construct counterexamples. But they are generally inductively strong, or, as I shall say, following Morreau (1997), *allowed.* Constitutive principles imply epistemic principles, but defeasibly inductively or nonmonotonically rather than deductively.

Consider a moral principle such as (4)b:

(4)   b.   Promises ought to be kept.

Ordinarily we would read this as a constitutive principle, saying that promises ought to be kept precisely because they are promises; their status as promises is responsible for their moral force. I shall signal the constitutive reading by inserting the phrase 'as such' (or, in conditionals, '*ipso facto*'), e.g., 'Promises, as such, ought to be kept'. To symbolize constitutive principles, I shall introduce a binary conditional connective $\gg$ to the language of standard deontic logic with a unary obligation operator O; (4)b emerges, at least initially, as

(8) $\forall x(\text{promise } x \gg O \text{ kept } x)$.

(As the phrase 'as such' suggests, there is a close link between constitutive principles as I shall construe them and Aristotle's use of *qua* phrases. The correspondences are not exact, however; see Bonevac forthcoming.)

One is not forced to read principle (4)b constitutively. A utilitarian, for example, might read (4)b epistemically, as a generalization based on a vast range of experience about good action, good character, and the good life. One could, in other words, read (4)b not as a *ceteris paribus* principle about *prima facie* obligation but as a generalization about actual obligations. I shall signal this by inserting the term 'generally'. "Promises generally ought to be kept" is an epistemic principle, claiming that promises are good epistemic indicators of actual obligations. To symbolize epistemic principles, I shall use Asher and Morreau's conditional connective $>$. (4)b, read epistemically, emerges as

(9) $\forall x(\text{promise } x > O \text{ kept } x)$.

In general, a principle of the form

(10) If A, then it ought to be the case that B

may be read in two ways. On a constitutive reading $(A \gg OB)$, it asserts that its being the case that A would, all other things being equal, make it obligatory that B. A's being the case, in other words, would be a reason for the *prima facie* obligation that B. On an epistemic reading $(A > OB)$, it asserts that, where A holds, there is usually — generally, normally, typically, etc. — an actual obligation that B.

The constitutive reading of moral principles tends to imply the epistemic reading. Consider the argument:

(11)   Promises, as such, ought to be kept.
   Therefore, promises generally ought to be kept.

Is this valid? No; it might be that promises as such ought to be kept in the sense that making a promise gives rise to a *prima facie* obligation to keep it, while also being the case that many, even most, promises ought not to be kept. It might happen, in other words, that being the fulfillment of a promise is a right-making but frequently or even generally overridden feature of an act. Consider a world in which promises are made only as acts of desperation. In such a world, promises would still have moral force. But so many promises might be made in mitigating circumstances, and with other important and morally relevant considerations at stake, that, from an epistemic point of view, promising is a poor indicator of actual obligation. At the opposite extreme, consider a world where people are so profligate with promises that they routinely commit themselves to incompatible courses of action. Again, promising would be a poor guide to actual obligation.

Nevertheless, (11) strikes me as reasonable. If promises, as such, ought to be kept, then it seems reasonable to expect that promises generally give rise to actual obligations. Someone who identifies promising as, *ceteris paribus*, inducing obligations does not do something bizarre by seeking to identify and track promises in the process of moral reasoning. Indeed, such a course of action seems entirely reasonable. Someone who maintains the truth of a constitutive principle quite naturally tends to use that principle in moral reasoning, not only in particular cases which would be intelligible even if (11) were rejected outright but also in developing a conception of how moral reasoning ought to proceed. A Kantian, for example, who maintains that promises as such ought to be kept may advise us to ask, of a prospective course of action, whether it involves the fulfillment or violation of a promise. That seems an entirely reasonable thing to do. But that can be so only if (11) is reasonable. The Kantian moves from a principle asserting that promising has moral force to recognizing promises as epistemic indicators of obligation, precisely the move made explicit in (11).

To give a more precise argument for the reasonableness of (11), I must develop further the notion of reasonableness itself. The key idea (from Morreau 1997) is that some worlds are more regular than others, in the sense that they involve fewer exceptions to principles, or conflicts between principles, than others. Say that world w is *constitutively irregular* with respect to A iff, for some B, $(A \gg B) \longrightarrow (A \longrightarrow B)$ is false at w, and *epistemically irregular* with respect to A iff, for some B, $(A > B) \longrightarrow (A \longrightarrow B)$ is false at w. World w is *constitutively (epistemically) irregular*, simpliciter, iff w is constitutively (epistemically) irregular with respect to some A. World w is as *constitutively (epistemically) regular* as w' if, whenever w is constitutively (epistemically) irregular with respect to A, so is w'; w is *more constitutively (epistemically) regular* than w' if w is as constitutively (epistemically) regular as w' but not vice versa. World w is *irregular* with respect to A if it is constitutively or epistemically irregular with respect to A, and *irregular* simpliciter if it is irregular with respect to some A. World w is as *regular as* w' if it is as regular as w' both constitutively and epistemically.

A world is a counterexample to an argument if, in that world, the premises are true and the conclusion is false. A counterexample to an argument is *unnecessarily irregular* if it contains irregularities not required by the truth of the argument's premises; if, that is, there are more regular worlds in which the premises and conclusion are all true. An argument is valid if it has no counterexamples, and *allowed* if all counterexamples to it are unnecessarily irregular. All valid arguments are allowed, vacuously, but not all allowed arguments are valid. If an argument is allowed, its premises *defeasibly imply* its conclusion.

This is enough to see somewhat more precisely why we may infer epistemic principles from constitutive principles. Argument (11) is not valid, but it is allowed, for all counterexamples to it are unnecessarily irregular. They contain conflicts that are not required by the truth of the premises. In describing a counterexample above, for instance, I said that there could be a world in which promises had moral force, justifying the constitutive reading of (4)b, but in which promises were routinely outweighed by other moral factors. That would be a world in which promises came into conflict with other moral considerations. Alternatively, I said, imagine a world in which so much promising was going on that people frequently faced choices between conflicting promises. These counterexamples rely on conflict that is not postulated by the argument's premises. They are therefore unnecessarily irregular.

3. Factual Principles

Nothing about the distinction between constitutive and epistemic principles is distinctively deontic. Consider the principle

(2) b. Humans are rational.

This principle, too, can be read as constitutive or epistemic. Most naturally, perhaps, we read it as constitutive, as asserting, that is, that humans are by rational by virtue of being human. But we might also read it epistemically, as asserting that humans are generally rational. On the former reading, (2)b asserts that humanity is productive of rationality. On the latter, it asserts that humanity is a good rationality indicator. We might represent these two readings as

(12) a. $\forall x(\text{human } x \gg \text{rational } x)$
b. $\forall x(\text{human } x > \text{rational } x)$.

A similar point applies to conditionals. Aristotle's example,

(7) b. If x cuts y's throat, x kills y.

may be read constitutively, as he intended; cutting y's throat is an effective means of killing y. But it may also be read epistemically; throat-cuttings are also good indicators of impending death. We might symbolize these readings as:

(13)   a. $\forall x \forall y(x$ cuts y's throat $\gg x$ kills $y)$
           b. $\forall x \forall y(x$ cuts y's throat $> x$ kills $y)$.

As with deontic principles, we may infer epistemic principles from corresponding constitutive principles. The following arguments are allowed, even though they are not deductively valid:

(14)   a.   Humans, as such, are rational.
               Therefore, humans are generally rational.
           b.   If x cuts y's throat, *ipso facto* x kills y
               Therefore, if x cuts y's throat, generally x kills y.

To see that (14)a is not deductively valid, imagine that humans, as such, are rational, in that part of the constitution of a human being is an apparatus that not only permits but tends to promote rational thought, but that the apparatus is so routinely circumvented or corrupted that rational thought and behavior are in fact quite rare. In such a world — where humans are more or less as they are now, but in which insanity is commonplace — (14)a's premise is true and its conclusion is false.

To see the same for (14)b, imagine that cutting a throat is, as in our world, an effective way of killing, but that nevertheless people and animals whose throats have been cut frequently do not die — because, say, throat-cutters are often incompetent, or because ubiquitous emergency teams have become very good at getting to the scene of throat-cuttings and patching up the victims. In such a world, it could still be true that, in the absence of other factors, throat cutting is fatal, but that throat cutting is not generally fatal because other factors are not generally absent.

Still, (14)a and b are allowed. Counterexamples to both are unnecessarily irregular. Consider (14)a: Counterexamples must be ones in which humans are as such rational but nevertheless not generally rational, because other factors intervene. Similarly with (14)b: Counterexamples are worlds in which throat cutting kills in the absence of other factors but in which other factors frequently intervene. In both cases, counterexamples involve conflicts among causal factors that are not required by the premises. Both arguments are therefore allowed.

To hold that arguments of the form

(15) $\forall x(Fx \gg Gx)$
           $\therefore \ \forall x(Fx > Gx)$

are allowed is to maintain that all counterexamples to such arguments are unnecessarily irregular, that is, that all counterexamples involve modus ponens failures not required by the premise. Suppose that the conclusion of (15) is false. Then, intuitively, there must be some normal (typical, standard) objects a such that Fa but not Ga. The premise of (15), however, implies that $Fa \gg Ga$. Any counterexample to (15), then, is a world in which, for some normal a,

(16) $\quad (Fa \gg Ga) \longrightarrow (Fa \longrightarrow Ga)$

is false. But that means that any counterexample to (15) is constitutively irregular.

So far I have been speaking as if all modus ponens failures result from the interference of other constitutive factors. That is, I have been speaking as if any counterexample to (15) — in which, for some a, (16) is false — is one in which another constitutive factor has overridden the premise of (15). Indeed, it is natural to assume, in the macroscopic realm, at any rate, that, where (16) fails, there is a property H such that

(17) $\quad \forall x(Hx \gg \neg Gx)\&Ha.$

(In one sense this is trivial; $\forall x(\neg Gx \gg \neg Gx)$ is valid in the system I shall present, so $\neg G$ itself constitutes an appropriate H. To make the assumption interesting, we must assume that $\neg \forall x \Box (Hx \to \neg Gx)$.) Suppose that humans as such are rational, but that Jones is irrational. It seems reasonable to assume that there is a property P such that Jones is irrational because Jones has P. Similarly, suppose that chairs as such have backs, but that this chair has no back. It seems reasonable to assume that there is a property this chair instantiates that is responsible for its having no back.

It seems to me somewhat less compelling to adopt an analogous assumption for epistemic principles. Suppose, for example, that world w is epistemically irregular. There are then some a, F, and G such that

(18) $\quad (Fa > Ga) \longrightarrow (Fa \longrightarrow Ga)$

is false. Does it follow that, for some H (such that $\neg \forall x \Box (Hx \to \neg Gx)$), either of the following holds?

(19)   a. $\forall x(Hx \gg \neg Gx)\&Ha$
          b. $\forall x(Hx > \neg Gx)\&Ha$

Consider some examples. Suppose that humans are generally omnivorous, but that Sarah is a vegetarian. Is there a property P in virtue of which Sarah is not omnivorous? (Note: being vegetarian will not itself fit the

bill, because vegetarians are of necessity not omnivorous.) Or suppose
that chairs generally have four legs, but that this chair has only one. Is
there a property it instantiates that is responsible for its having one leg?
I am not sure. Generally, in a game of poker, nobody gets a royal flush.
But it does not seem to follow that a poker player who is lucky enough to
hold one does so in virtue of instantiating some other property.

4. Truth Conditions for Epistemic Principles

I shall present a formal system for epistemic principles using the Asher-
Morreau system of commonsense entailment. The key, of course, lies
in the truth conditions for epistemic principles. When is an epistemic
principle $A > B$ true at a world w? David Lewis has given an influential
and plausible account of counterfactual conditionals, according to which
(for the sake of simplicity, adopting the limit assumption) a conditional
$A \square \to B$ is true at a world w iff B is true at all the closest A-worlds, i.e.,
the worlds most similar to w in which A is true (Lewis 1973). Where A is
a sentence, let $[A]$ be the proposition A expresses, i.e., the set of worlds in
which A is true. Expressed in terms of a selection function f from worlds
w and sentences A (or, given Lewis's theory, the proposition $[A]$) to sets
of worlds $f(w, [A])$ (intuitively, the closest A-worlds to w), we can express
the Lewis truth condition as

(20) $A \square \to B$ is true at w iff $f(w, [A]) \subseteq [B]$.

The Asher-Morreau truth condition is almost identical:

(21) $A > B$ is true at w iff $*(w, [A]) \subseteq [B]$.

Both truth conditions have the form Chellas (1974, 1980) specifies as the
paradigm for conditionals in general. $*(w, [A])$ is the set of A-worlds in
which conditions are suitable for assessing (relative to w) what happens
when A. According to this truth condition, 'If A, then generally B' is true
at world w iff B is true in all A-worlds in which conditions are suitable for
assessing (relative to w) what happens when A.[3] The differences between

---

[3]Morreau (1997) suggests a slight revision for fainthearted conditionals:

(21′) 'If A, then (provided that conditions are suitable for a contextually determined
end e) B' is true at world w if and only B is true in the closest A-worlds to w in
which conditions are suitable (relative to w) for assessing what happens when
A given e.

This differs from the Asher-Morreau truth condition by (a) considering only the closest
p-worlds in which conditions are suitable and (b) incorporating a contextually determined end, which sometimes depends on the consequent q. Letting '$A >_e B$' represent

the Asher-Morreau and Lewis truth conditions lie solely in the constraints imposed on the selection function. The most important difference is that the Asher-Morreau function is not centered; there is no requirement that w belong to $*(w, [A])$. In w, conditions may not be suitable for assessing what happens when A. So, while one can think of $*(w, [A])$ as the set of epistemically normal or typical A-worlds, one cannot think of it as the set of *closest* A-worlds. Any world is the closest world to itself, but it is not necessarily normal or typical even by its own standards.

Let L be a language with the usual first-order operators, the unary obligation operator O, the unary necessity operator $\Box$, and an epistemic conditional $>$. All symbols are names of themselves. Throughout the following, take w, w', etc. to be arbitrary worlds in a nonempty set W and take p, q, etc. to be arbitrary subsets of W. $\oplus w$ is the set of w's ideal worlds; $\oplus(X)$, where X is a set of worlds, is $\cup \oplus (w)$ for $w \in X$; $*(w, p)$ is the set of w-p-normal worlds. Note that

(23)   a. $p \subseteq q \to \oplus p \subseteq \oplus q$
        b. $\oplus(p \cup q) = \oplus p \cup \oplus q$
        c. $\oplus(p \cap q) \subseteq \oplus p \cap \oplus q$

To preview truth conditions: A unary obligation statement OA is true in w if A holds in all ideal worlds, i.e., if $\oplus(w) \subseteq [A]$. A necessity statement $\Box A$ is true in w if A holds in all $w \in W$. (The logic of $\Box$ is thus S5.) A generic $A > B$ is true in w if B is true in all normal worlds relative to A and w; if, in other words, $*(w, [A]) \subseteq [B]$. It follows that an epistemic principle $A > OB$ is true in w if $\oplus * (w, [A]) \subset [B]$, and that $\Box(A \to B)$ entails $A > B$.

Definition. An L *frame* $\Im$ is a quadruple $\langle W, D, \oplus, * \rangle$, where W is a nonempty set of worlds, D a nonempty set of individuals, $\oplus$ a function from W to $\wp(W)$, and $*$ a function from $W \times \wp(W)$ to $\wp(W)$.

'If A, then (provided that conditions are suitable for end e) B', we can write this truth condition in symbols using the concept of a selection function:

(21″) $A >_e B$ is true at w if and only if $f(w, [A], e) \subseteq [b]$.

Here, $f(w, [A], e)$ is the set of closest A-worlds to w in which conditions are suitable (relative to w) for assessing what happens when A given a (contextually determined) end e. (The function f is thus the composition of a Lewis selection function yielding the closest worlds and the $*$ function of the Asher-Morreau condition expressed in (21) with an extra argument place for end e.)

Adding the contextually determined ends to the truth conditions is useful for dealing with the drowning problem (Benerhat *et al.* 1993), which is beyond the scope of this paper. I have not added them here, partly for the sake of simplicity, and partly because I am not entirely convinced that the drowning problem ought to be solved; see Bonevac (1998).

Definition. *Proper L* frames are frames obeying:

(a) seriality: $\oplus(w) \neq \emptyset$;

(b) facticity: $*(w, p) \subseteq p$; and

(c) disjunction: $*(w, p \cup q) \subseteq *(w, p) \cup *(w, q)$.

Seriality guarantees that O is an actual obligation operator, admitting no moral dilemmas. ($\oplus p = \emptyset \rightarrow p = \emptyset$). Facticity stipulates that p is one of the things that normally hold when p holds, making sentences such as 'Humans are human' valid. Disjunction guarantees the validity in L of inferences such as

(21)   If x cuts y's throat, x kills y.
         If x breaks y's neck, x kills y.
         ———————————————————————
         If x cuts y's throat or breaks y's neck, x kills y.

As Morreau (1992) shows, facticity and disjunction entail specificity:

(22) $(p \subseteq q \,\&\, *(w, p) \cap *(w, q) = \emptyset) \Rightarrow *(w, q) \cap p = \emptyset$.[4]

Definition. A *base model* is a tuple $\langle W, D, \oplus, *, [\,] \rangle$, where $\langle W, D, \oplus, * \rangle$ is a proper L frame and $[\,]$ is a function from nonlogical constants of L to intensions (which, in turn, are functions from worlds to extensions).

   A variable assignment α is a function from variables into D, the domain of a base model; for the purposes of this paper, all worlds have the same domain. Assume that all objects in D have names in L, and assume bivalence. Truth is satisfaction under all assignments: $M, w \models A$ iff $\forall \alpha M, w, \alpha \models A$. Where Γ is a set of formulas, $M, w, \alpha \models \Gamma$ iff $M, w, \alpha \models A$ for all $A \in \Gamma$. The interpretation function $[\,]$ extends to formulas with the following definitions. $[A]_{M,\alpha}$, the proposition A expresses under α in M, is the set of worlds in which α satisfies $A : \{w \in W : M, w, \alpha \models A\}$. If A is a sentence, and Γ a set thereof, we write $[A]_M$ for $\{w \in W : M, w \models A\}$ and $[\Gamma]_M$ for $\cap\{[A]_M : A \in \Gamma\}$.

Definition.

   $\delta(t) = [t]$, if t is a constant, and $\alpha(t)$, if t is a variable

---

[4]Proof: Assume facticity and disjunction. Assume further that $p \subseteq q \,\&\, *(w, p) \cap *(w, q) = \emptyset$. Since $p \subseteq q$, $p \cap q = p$. So $q = p \cup (q - p)$. By Disjunction, $*(w, q) \subseteq *(w, p) \cup *(w, q - p)$. Since $*(w, q) \cap *(w, p) = \emptyset$, $*(w, q) \subseteq *(w, q - p)$. But, by facticity, $*(w, q - p) \subseteq q - p \subseteq -p$. It follows that $*(w, q) \cap p = \emptyset$.

$M, w, \alpha \models Rx_1 \ldots x_n$ iff $\langle \delta(x_1), \ldots, \delta(x_n) \rangle \in [R]_{M,w}$

$M, w, \alpha \models \neg A$ iff not $M, w, \alpha \models A$

$M, w, \alpha \models A\&B$ iff $M, w, \alpha \models A$ and $M, w, \alpha \models B$

$M, w, \alpha \models \forall x A$ iff $M, w, \alpha' \models A$, for all α' such that, for all $v \neq x$, $\alpha(v) = \alpha'(v)$

$M, w, \alpha \models OA$ iff $\oplus(w) \subseteq [A]_{M,\alpha}$

$M, w, \alpha \models \Box A$ iff $W = [A]_{M,\alpha}$

$M, w, \alpha \models A > B$ iff $*(w, [A]_{M,\alpha}) \subseteq [B]_{M,\alpha}$

A generic conditional $A > B$ holds in w if B holds in all worlds normal with respect to w and A. Abbreviating, B must hold in all the w-A-normal worlds.

Definition. If A is an L formula, and Γ a set thereof, then $\Gamma \models A$ iff, for all M, w, and α, if $M, w, \alpha \models \Gamma$, then $M, w, \alpha \models A$. A is *valid*, $\models A$, iff $\emptyset \models A$.

The following constitute a sound and complete axiom system for commonsense entailment:

(i) Truth-functional L-tautologies, modus ponens (for $\rightarrow$), and replacement

(ii) Standard axioms and rules for the quantifiers:

$\forall x A \rightarrow A(t/x)$ for any term t

$\forall x\, A \leftrightarrow \neg \exists x \neg A x$

$\forall x (A \rightarrow B) \rightarrow (\exists x A \rightarrow B)$ for x not free in B

$\vdash A \rightarrow B(t/x) \Rightarrow \vdash A \rightarrow \forall x B$, where t is a constant not in A or B

$\vdash A \Rightarrow \vdash A(t/x)$ where t is a term not in A

(iii) Standard axioms and rules for necessity:

$\Box(A \rightarrow B) \rightarrow (\Box A \rightarrow \Box B)$

$\Box A \rightarrow A$

$\Box A \rightarrow \Box\Box A$

$\neg\Box A \rightarrow \Box\neg\Box A$

$\vdash A \Rightarrow \vdash \Box A$

(iv) Standard axioms and rules for obligation:

$$\Box A \to OA$$
$$O(A \to B) \to (OA \to OB)$$
$$OA \to \neg\Box\neg A$$

(v) Axioms and rules for epistemic conditionals:

$$A > A$$
$$((A > C)\&(B > C)) \to ((A \lor B) > C)$$
$$\forall x(A > B) \to (A > \forall x B), \text{ for } x \text{ not free in } A.$$
$$\vdash (B_1 \& \dots \& B_i) \to B \Rightarrow \vdash (A > B_1 \& \dots \& A > B_i) \to A > B$$
$$\vdash A \leftrightarrow B \Rightarrow \vdash A > C \leftrightarrow B > C$$

Epistemic conditionals are closed under logical consequence. Epistemic conditionals with equivalent antecedents, moreover, are equivalent. Argument patterns that fail for counterfactuals — strengthening the antecedent, transitivity, and contraposition — typically fail for epistemic conditionals.

A valid pattern worth remarking is what, in deontic contexts, Bernard Williams (1973) calls agglomeration and Ruth Barcan Marcus (1980) calls factoring:

(23)   Promises ought to be remembered.
Promises ought to be kept.
Therefore, promises ought to be remembered and kept.

In nondeontic contexts, this becomes

(24) $A > B$
$A > C$
$\therefore A > (B\&C)$

for example,

(25)   Humans are omnivorous.
Humans are weak relative to other large animals.
Therefore, humans are omnivorous but weak relative to other
large animals.

Examining these and other natural language arguments of similar forms lends support to the idea that (24) holds for epistemic conditionals.

This has an important consequence: the 'generally' with which I have been marking epistemic generics and conditionals should not be understood in probabilistic terms. If $A > B$ meant something like 'the conditional probability of B on A is high', (24) would be invalid. Suppose someone were to propose the truth condition:

(26) $A > B$ is true in w iff $P(B/A) > n$

for some suitable value of n. Then, from $P(B/A) > n$ and $P(C/A) > n$ we could infer $P(B\&C/A) > n$ only if $P(B/A) = 1$ or $P(C/A) = 1$. That is, a probabilistic analysis would validate not (24) but the related pattern:

(27) $A > B$
$\quad\ \Box(A \to C)$
$\quad\ \therefore\ A > (B\&C)$

If arguments such as (25) are valid — as I believe they are — then a probabilistic analysis of this kind is inadequate.

More interesting than the valid patterns are the allowed patterns. I shall focus only on four argument forms, three of which are allowed.

Defeasible modus ponens. Modus ponens is not valid for epistemic conditionals, but it is allowed. These, for example, are reasonable even if not deductively valid:

(28) a. Chairs have four legs. $\qquad\qquad\ \forall x(Cx > Fx)$
$\qquad\quad$ That is a chair. $\qquad\qquad\qquad\ Ca$
$\qquad\quad$ Therefore, that has four legs. $\quad\ \therefore\ Fa$
$\qquad$ b. If you promised to go, you should go. $\ p > Oq$
$\qquad\quad$ You promised to go. $\qquad\qquad\quad p$
$\qquad\quad$ Therefore, you should go. $\qquad\ \therefore\ Oq$

Counterexamples to these arguments are unnecessarily irregular. A counterexample to (28)a, for example, would have to be one in which $Ca > Fa$ and Ca were true, but Fa was false — in which, therefore, $(Ca > Fa) \longrightarrow (Ca \longrightarrow Fa)$ was false — making it epistemically irregular with respect to Ca. There are more regular worlds in which the premises and conclusion are all true (for example, any world in which $Ca > Fa$, Ca, and Fa are all true). So, any counterexample to (28)a would be unnecessarily irregular, making (28)a allowed. An analogous argument holds for (28)b.

This explains the usefulness of principles. As Dancy (1991) argues, they do have exceptions, and arguments from them to particular conclusions are not valid. They nevertheless function effectively in reasoning, for arguments from principles to particular conclusions may be allowed. From a deductive point of view, one might think that an infinite amount of information is required to move from a general principle to a particular conclusion; one would have to rule out all possible counterexamples. From a different point of view, however, one reasons given the assumption that the world is as normal as the premises allow it to be. One assumes, that is, that any reasons for abnormality or exceptions to principles have been mentioned. This is not certain, of course; additional information may undermine the conclusions reached by way of this assumption. But that is always the case with inductive reasoning.

Conflict. Generally, when the premises give conflicting information, no definite conclusions are allowed. Which information takes precedence is a substantive question that cannot be decided on logical grounds alone. These examples are neither valid nor allowed. (I use (!) to mark disallowed conclusions.)

(29)	a.	That is a chair.	Ca
		That does not have four legs.	$\neg Fa$
		Chairs have four legs.	$\forall x(Cx > Fx)$
		Therefore, that has four legs. (!)	$\therefore Fa(!)$
	b.	Quakers are pacifists.	$\forall x(Qx > Px)$
		Republicans are not pacifists.	$\forall x(Rx > \neg Px)$
		Nixon is a Republican and a Quaker.	$Rn\&Qn$
		Therefore, Nixon is a pacifist. (!)	$\therefore Pn(!)$
		Therefore, Nixon is not a pacifist. (!)	$\therefore \neg Pn(!)$
	c.	Judges should be partial to no one.	$\forall x(Jx > ONx)$
		Fathers should be partial to their children.	$\forall x(Fx > OCx)$
		Sam is a judge and a father.	$Js \& Fs$
		Therefore, Sam should be partial to his children. (!)	$\therefore OCs(!)$
		Therefore, Sam should be partial to no one. (!)	$\therefore ONs(!)$

To see why these conclusions are not allowed, consider (29)a. Any counterexample makes Ca, $\neg Fa$, and $Ca > Fa$ true. It thus makes $(Ca > Fa) \rightarrow (Ca \rightarrow Fa)$ false. Any counterexample, therefore, is irregular with respect to Ca. But is it unnecessarily irregular? No. Any world in which the premises are true has the same property. There are no worlds in which the premises and conclusion are all true that are any more regular (by, in particular, being regular with respect to Ca).

The same reasoning applies to (29)b and c. Consider just (29)b, the celebrated Nixon diamond. Any world making the premises true must be irregular with respect to either Qn or Rn. In any world making the premises true, that is, Nixon is either an atypical Quaker or an atypical Republican. The conclusion that Nixon is a pacifist is disallowed, for a counterexample would be irregular with respect to Qn (making Nixon, in other words, an atypical Quaker) and there would be no more regular worlds in which the premises and conclusion were all true. Any world satisfying the premises that is regular with respect to Qn is irregular with respect to Rn. The counterexample, then, is irregular but not unnecessarily so. The conclusion that Nixon is not a pacifist is also disallowed, for similar reasons. Any counterexample is irregular with respect to Rn, making Nixon an atypical Republican. A world making the premises all true and being regular with respect to Rn would be irregular with respect to Qn. So, there are no more regular worlds in which the premises and conclusion are all true.

Disjunctive Conflict. When the premises give conflicting information, one may draw disjunctive conclusions. They are not valid; there is no guarantee that there are not further conflicts that would complicate the situation beyond what is envisioned in the premises. But they are allowed; they are reasonable inferences from the information in the premises.

(30) a. Judges should be partial to no one. $\forall x(Jx > ONx)$
           Fathers should be partial to their
                             children. $\forall x(Fx > OCx)$
           Sam is a judge and a father. $Js\&Fs$
           One cannot be partial to no one and to
                        one's children. $\forall x\Box(Nx\&Cx)$
           Therefore, Sam should be partial to his
                        children or to no one. $\therefore ONs \lor OCs$
      b. Quakers are pacifists. $\forall x(Qx > Px)$
           Republicans are hawks. $\forall x(Rx > Hx)$
           Nixon is a Republican and a Quaker. $Rn\&Qn$
           No one could be a pacifist and a hawk. $\forall x\Box\neg(Px\&Hx)$
           Therefore, Nixon is either a pacifist or
           a hawk. $\therefore Pn \lor Hn$

Consider just (30)b. Any counterexample makes $Qn > Pn$, $Rn > Hn$, Rn, and Qn true but Pn and Hn false. It is thus irregular with respect to both Qn and Rn, for it makes both $(Qn > Pn) \longrightarrow (Qn \longrightarrow Pn)$ and $(Rn > Hn) \longrightarrow (Rn \longrightarrow Hn)$ false. It is thus unnecessarily irregular, for there are more regular worlds making all the premises true. Because being a pacifist is incompatible with being a hawk, no such worlds are regular with respect to both Qn and Rn. But worlds which are irregular with respect to just one can make the premises and conclusion all true.

Counterexamples to (30)b, in other words, make Nixon an atypical Quaker
and an atypical Republican. Given the premises, he must be one or the
other, but he need not be both. Any counterexample to (30)b is therefore
unnecessarily irregular. So, (30)b is allowed.[5]

Specificity. In general, conflicting information yields only disjunctive
conclusions. But, when the antecedents of the conflicting conditionals are
logically related so that one entails the other, more specific conclusions
are allowed. More specific information takes precedence over less specific
information. Consider:

(31)	a.	Quakers are pacifists.	$\forall x(Qx > Px)$
		Western Quakers are not pacifists.	$\forall x(Wx > \neg Px)$
		Necessarily, western Quakers are Quakers.	$\forall x \Box(Wx \rightarrow Qx)$
		Nixon is a western Quaker.	Wn
		Therefore, Nixon is not a pacifist.	$\therefore \neg Pn$
	b.	Promises should be kept.	$\forall x(Px > OKx)$
		Promises to do wrong should not be kept.	$x(Wx > O\neg Kx)$
		Promises to do wrong, necessarily, are promises.	$\forall x \Box(Wx \rightarrow Px)$
		Jack's promise was a promise to do wrong.	Wj
		Therefore, Jack's promise should not be kept.	$\therefore O\neg Kj$

To see that (31)a is allowed, think about an arbitrary counterexample w.
It would make $Wn > \neg Pn$, Wn, and Pn all true. It would thus falsify
$(Wn > \neg Pn) \rightarrow (Wn \rightarrow \neg Pn)$. So, w would be irregular with respect to
Wn. I now argue that w would be unnecessarily irregular. It might seem
that this is not so; by the reasoning used for (29)b, any world in which the
premises are all true that is regular with respect to Wn is irregular with
respect to Qn. But this does not imply that there are no more regular
worlds making the premises true, for w is already irregular with respect
to Qn. The premises already require that Nixon is an atypical Quaker.

Recall that facticity and disjunction imply specificity:

(22) $(p \subseteq q \ \& * (w,p) \cap *(w,q) = \emptyset) \Rightarrow *(w,q) \cap p = \emptyset.$

---

[5]If being a pacifist were compatible with being a hawk, then the conclusion that
Nixon is both a pacifist and a hawk would be allowed. Nothing in the premises would
then force any irregularity. This illustrates the nonmonotonicity of allowed arguments:
the first three premises of (30)b support the conclusion $Pn \ \& \ Hn$, but (30)b as a whole
does not.

Necessarily, all western Quakers are Quakers, so $[Wn] \subseteq [Qn]$. And western Quakers are not pacifists, but Quakers are, so $*(w, [Wn]) \subset [\neg Pn]$ and $*(w, [Qn]) \subset [Pn]$. Thus, $*(w, [Wn])$ and $*(w, [Qn])$ are disjoint. According to (22), so are $[Wn]$ and $*(w, [Qn])$. That is, Quakers are generally not western Quakers. So, w makes $Qn > \neg Wn$, Qn, and Wn all true, falsifying $(Qn > \neg Wn) \to (Qn \to \neg Wn)$. Counterexample w is thus already irregular with respect to Qn. Nixon, being a western Quaker, is already an atypical Quaker. A world in which the premises and conclusion of (31)a are all true is thus more regular than w, despite its irregularity with respect to Qn. So, w is unnecessarily irregular, and (31)a is allowed.

5. Truth Conditions for Constitutive Principles

How can we represent constitutive principles in this framework? Earlier, Asher and I (1996, 1997) in effect introduced another constitutive conditional connective, with truth conditions analogous to those for epistemic conditionals:

(32) $A \gg B$ is true at w iff $\bullet(w, [A]) \subseteq [B]$.

$\bullet$, like $*$, is a function from worlds and propositions to sets of worlds — intuitively, the constitutively normal or typical A-worlds, the worlds appropriate for assessing constitutively what happens when A.

To make such an approach work, one must introduce facticity and disjunction constraints for $\bullet$, add axioms and rules for $\gg$:

(vi) Axioms and rules for constitutive conditionals:

$A \gg A$

$((A \gg C)\&(B \gg C)) \to ((A \vee B) \gg C)$

$\forall x(A \gg B) \to (A \gg \forall xB)$, for x not free in A.

$\vdash (B_1 \& \ldots \& B_i) \to B \Rightarrow \vdash (A \gg B_1 \& \ldots \& A \gg B_i) \to A \gg B$

$\vdash A \leftrightarrow B \Rightarrow \vdash A \gg C \leftrightarrow B \gg C$

and somehow link the two selection functions so that arguments from constitutive to epistemic conditionals are allowed. One might, for example, simply introduce

(33) $(A \gg B) > (A > B)$

as an axiom, and add a corresponding semantic constraint. But the system that results is unpleasant. The duplication of axioms and constraints is bad enough; it strongly suggests that a useful generalization is being missed. Worse, such a system provides no explanation for why constitutive and epistemic principles relate as they do. The inference from constitutive to epistemic principles is imposed by fiat through (33), not explained by the semantic analysis of those principles.

I wish, instead, to adopt a very simple analysis of constitutive principles in terms of epistemic ones. To do that, it is useful to add $\top$ to the language and define an additional inner necessity operator:

(34) $M, w, \alpha \models \top$ for all M, w, and α

(35) $\boxdot A = \top > A$

Read as "normally," "generally," "typically," etc., $\boxdot$ is an inner modality in the sense that

(36) $\Box A \to \boxdot A$

but not conversely. The logic of this operator is the minimal normal modal logic K, for distribution and necessitation hold:

(37)   a. $\boxdot(A \to B) \to (\boxdot A \to \boxdot B)$
     b. $\vdash A \Rightarrow \vdash \boxdot A$

Say that w is normal relative to w' (Nww') if $w \in *(w', W)$. Then the following hold under the conditions listed:

(38)   a. $\boxdot A \to A$ iff N is reflexive
     b. $\boxdot A \to \boxdot\boxdot A$ iff N is transitive
     c. $\neg\boxdot A \to \boxdot\neg\boxdot A$ iff N is symmetric

In general, however, none of these conditions on N can be expected to hold.

Still, there is something to be said for one of them. (38)a might be expressed as an argument form

(39) $\boxdot A$
     $\therefore A$

This is not valid, but it is allowed. Any counterexample makes $\boxdot A$ true and A false. It thus makes $(\top > A) \to (\top \to A)$ false, and, so, is irregular with respect to $\top$. The irregularity is unnecessary, for there are more regular worlds in which premise and conclusion are both true.[6]

---

[6]Note that it does not follow that $\boxdot A > A$ is valid. It is not even allowed. Let $\models\!\approx$ represent the entailment relation corresponding to allowed arguments. Then, if $\models\!\approx A > B$, $A \models\!\approx B$. This is simply defeasible modus ponens. But the converse does not hold;

I can now state my proposal for constitutive conditionals:

(40) $A \gg B = \Box(A > B)$,

or, in English,

(41) If A, then *ipso facto* B = Generally, if A, then generally B.

This has the desired effect of allowing defeasible inferences from constitutive to epistemic principles:[7]

(42) a. $A \gg B$
        $\therefore A > B$
     b. $\forall x(Fx \gg Gx)$
        $\therefore \forall x(Fx > Gx)$

It also yields as theorems concerning constitutive conditionals the analogues to the axioms and rules for epistemic conditionals, given as (vi) above:

(43) a. $A \gg A$
     b. $((A \gg C)\&(B \gg C)) \rightarrow ((A \lor B) \gg C)$
     c. $\forall x(A \gg B) \rightarrow (A \gg \forall xB)$, for x not free in A.
     d. $\vdash (B_1 \& \ldots \& B_i) \rightarrow B \Rightarrow \vdash (A \gg B_1 \& \ldots \& A \gg B_i)$
        $\rightarrow A \gg B$
     e. $\vdash A \leftrightarrow B \Rightarrow \vdash A \gg C \leftrightarrow B \gg C$

Again, the allowed inferences are more interesting than the valid ones. The patterns allowed and disallowed for constitutive principles mimic those allowed and disallowed for epistemic principles.

Defeasible modus ponens. Modus ponens is allowed for constitutive conditionals despite its invalidity:

---

allowed arguments and epistemic conditionals are not linked by conditionalization. For the same reason, (33) implies that the argument forms of (42) are allowed, but not conversely.

    [7]This is the chief argument for characterizing $A \gg B$ as $\Box(A > B)$ rather than $\Box (A > B)$; the latter would make (42)a and b valid rather than allowed. As I have argued above, however, constitutive principles imply epistemic principles, not outright, but defeasibly.

    In addition, defining constitutive principles in terms of necessity rather than inner necessity would, given a logic for necessity at least as strong as S4, make constitutive principles themselves necessary. On the view advanced in this paper, however, such principles may be contingent.

(44)   a.   Chairs have backs. $\forall x(Cx \gg Bx)$
            That is a chair. Ca
            Therefore, that has a back. $\therefore Ba$
       b.   If you promised to go, you should go. $p \gg Oq$
            You promised to go. p
            Therefore, you should go. $\therefore Oq$

Conflict. Generally, when the premises involve conflicting information, of a constitutive or epistemic nature, no definite conclusions are allowed.

(45)   a.   That is a chair. Ca
            That does not have a back. $\neg Ba$
            Chairs have backs. $\forall x(Cx \gg Bx)$
            Therefore, that has a back. (!) $\therefore Ba$ (!)
       b.   Quakers are pacifists. $\forall x(Qx \gg Px)$
            Republicans are not pacifists. $\forall x(Rx > \neg Px)$
            Nixon is a Republican and a Quaker. $Rn\&Qn$
            Therefore, Nixon is a pacifist. (!) $\therefore Pn$ (!)
            Therefore, Nixon is not a pacifist. (!) $\therefore \neg Pn$ (!)
       c.   Judges should be partial to no one. $\forall x(Jx \gg ONx)$
            Fathers should be partial to their
                               children. $\forall x(Fx \gg OCx)$
            Sam is a judge and a father. $Js\&Fs$
            Therefore, Sam should be partial to
                               his children. (!) $\therefore OCs$ (!)
            Therefore, Sam should be partial to
                               no one. (!) $\therefore ONs$ (!)

It does not matter whether the principles involved are constitutive or epistemic. One might think that, in cases of conflict, constitutive principles take precedence over epistemic principles (they are, after all, defeasibly stronger) or, conversely, one might think that epistemic principles should take precedence over constitutive principles (who cares what is constitutive if it is overridden for epistemic purposes?). But consider (45)b. It seems plausible to think that the principle that Quakers are pacifists is constitutive — being a pacifist is one of the cluster of beliefs involved in being a Quaker — while the principle that Republicans are not pacifists is epistemic, a good indicator at present but not tied to what it is to be a Republican. (If this seems implausible, think of the Republican isolationists of the 1920s and 1930s.) Now, you learn that Nixon (about whom, let's say, you have no other information) is both a Republican and a Quaker. What are you allowed to conclude? Intuitions may differ, but it seems to me that the right answer is nothing. Given this information alone, I find it slightly more likely that Nixon is a pacifist. This could be a slight preference for constitutive principles, though I assess the situation similarly if both principles are read as constitutive or as epistemic. In any case, the difference is slight, not enough to justify even a defeasible conclusion.

My theory agrees. One cannot conclude that Nixon is a pacifist; any counterexample is irregular with respect to Qn. But it is not unnecessarily irregular, for worlds making Pn true are irregular with respect to either Rn or $\top$. In such worlds, Rn, Pn, $\top > (Rn > \neg Pn)$, and of course $\top$ are all true. Now $Rn > \neg Pn$ is either true or false. If it is true, then the world is irregular with respect to Rn, for $(Rn > \neg Pn) \to (Rn \to \neg Pn)$ fails. If it is false, then the world is irregular with respect to $\top$, for $(\top > (Rn > \neg Pn)) \to (\top \to (Rn > \neg Pn))$ fails.

Disjunctive Conflict. When premises give conflicting information, one may draw disjunctive conclusions — again, whether the conflict occurs solely among constitutive principles, solely among epistemic principles, or between the two:

(46)	a.	Judges should be partial to no one.	$\forall x(Jx \gg ONx)$
		Fathers should be partial to their children.	$\forall x(Fx \gg OCx)$
		Sam is a judge and a father.	$Js\&Fs$
		One cannot be partial to no one and to one's children.	$\forall x \Box \neg(Nx\&Cx)$
		Therefore, Sam should be partial to his children or to no one.	$\therefore\ ONs \lor OCs$
	b.	Quakers are pacifists.	$\forall x(Qx \gg Px)$
		Republicans are hawks.	$\forall x(Rx > Hx)$
		Nixon is a Republican and a Quaker.	$Rn\&Qn$
		No one could be a pacifist and a hawk.	$\forall x \Box \neg(Px\&Hx)$
		Therefore, Nixon is either a pacifist or a hawk.	$\therefore\ Pn \lor Hn$

Specificity. When the antecedents of the conflicting conditionals are logically related so that one entails the other, more specific information takes precedence over less specific information, no matter the status of the principles involved. Consider:

(47)	a.	Quakers are pacifists.	$\forall x(Qx \gg Px)$
		Western Quakers are not pacifists.	$\forall x(Wx > \neg Px)$
		Necessarily, western Quakers are Quakers.	$\forall x \Box(Wx \to Qx)$
		Nixon is a western Quaker.	Wn

	Therefore, Nixon is not a pacifist.	$\therefore \neg Pn$
b.	Promises should be kept.	$\forall x(Px > OKx)$
	Promises to do wrong should not be kept.	$\forall x(Wx \gg O\neg Kx)$
	Promises to do wrong, necessarily, are promises.	$\forall x \Box(Wx \to Px)$
	Jack's promise was a promise to do wrong.	Wj
	Therefore, Jack's promise should not be kept.	$\therefore O\neg Kj$

We can see, then, why the distinction between constitutive and epistemic principles is generally not marked in discourse. Logically, it makes no difference which principles are which. The logic of constitutive principles is just the logic of epistemic principles. Ordinarily, they are interchangeable. The only difference is that constitutive principles are defeasibly stronger than epistemic principles.

Logically, then, defining constitutive principles in terms of corresponding epistemic conditionals is quite satisfactory. But is it philosophically satisfactory? It runs contrary to the intuition that constitutive principles are more fundamental than epistemic principles. Constitutive principles, one might think, pertain to what things are; epistemic principles pertain to what we can infer things to be. The former seems primary, the latter, derivative.

But that is not the only way to look at things. The inner modality $\boxdot$ can fairly be read as "generally," "typically," or "normally." But it is in some ways stronger than these locutions suggest. If $\boxdot A$ is true in w, then A is true in every world in which conditions are suitable for assessing (relative to w) what happens when the laws of logic hold, i.e., for assessing much of anything. $\boxdot A$ is true in w if $*(w, W) \subseteq [A]$, and $*(w, W)$ is a large set. Intuitively, it is the set of epistemically typical or normal worlds, the worlds in which everything that is generally true *no matter what* is true. Think of these as worlds that share the fundamental laws of w and also have contingent initial conditions that are not too bizarre from w's perspective. Sentences true in all such worlds are general epistemic defaults. They have exceptions, but, from the perspective of w, they are generally true no matter what the particular facts are. One may, therefore, interpret $\boxdot A$ as saying that A is *lawlike*, and think of $*(w, W)$ as the set of *lawlike worlds* relative to w.[8]

---

[8]This does not entail that lawlike statements are necessary, or even true. $\boxdot A$ does not entail A, $\Box A$, $\boxdot\boxdot A$, or $\Box \boxdot A$. ($\boxdot A$ does, however, defeasibly imply A.) Some may think that this makes the adjective 'lawlike' inappropriate, on the grounds that laws are necessary. But the sense of 'lawlike' I am using not that applicable to fundamental

From this perspective, (40) is not so surprising. It says that constitutive principles are lawlike epistemic principles. It says, in other words, that a constitutive conditional $A \gg B$ holds if the corresponding epistemic conditional $A > B$ is lawlike.

Consider some of the constitutive-epistemic principle pairs with which I began this paper:

(2)   b.   Humans are rational.
        c.   Humans are omnivorous.

What is the difference? According to the analysis I am proposing, (2)b differs from (2)c in that the generalization it makes is lawlike. It is not just that, in this world, being human is a good indicator of being rational; it is that that would be true in a wide range of worlds, the worlds that share our laws and exhibit contingent circumstances that are not too bizarre from our point of view.

(4)   b.   Promises ought to be kept.
        c.   Promises are easily forgotten.

Principle (4)b defeasibly implies that promising is a good indicator of obligation, but implies furthermore that the connection between promising and obligation is lawlike. Not only is promising a good indicator of obligation in our world; it is *generally* a good indicator of obligation. That is, it is a good indicator in any world that shares the general shape of our world.

(5)   b.   If that is an insect, it has six legs.
        c.   If that is an insect, it has wings.

(7)   b.   If x cuts y's throat, x kills y.
        c.   If x cuts y's throat, x intends to kill y.

Similar points may be made about these examples. The link between being an insect and having six legs is lawlike in a way that the link between being an insect and having wings is not. And there is a lawlike connection between cutting an animal's throat and killing it. That is not true of the epistemic principles.

---

laws of nature, which may be necessary and admit no exceptions — I take no position on those issues in this paper — but rather to generalizations such as 'What goes up must come down', 'Insects have six legs', 'Birds fly', etc. These are constitutive: they are not accidental, or even solely epistemic, but they are not necessary either. It is appropriate to think of them as laws, but as derivative rather than basic laws, which admit exceptions because of the complex interactions of fundamental laws, initial conditions, and resulting circumstances.

These examples suggest a more general point. Epistemic principles depend on merely distributional aspects of the actual world in a way that constitutive principles do not. (5)b is true in our world, but it would be false in a world that differed from ours, not in fundamental laws or the general shape of initial conditions and resulting circumstances, but simply in distribution: in having fleas, ticks, and lice, for example, be vastly more common than flies, bees, or beetles. Similarly, (7)c is true in our world, if y ranges over nonhuman animals, but could easily be false without anything other than distributional changes: an increase in vegetarianism, for example. The same, obviously, is true of (2)c. And (4)c could be false if the distribution between serious and commonplace promises were to shift in favor of the former.

6. Conclusion

I have argued that Aristotle's account in *Metaphysics* Δ implies that there are three kinds of essential properties: those that inhere in their subjects (a) of necessity, (b) because of what their subjects are, and (c) usually. The first kind has been well-studied, and is well-represented by the strict conditional. The second corresponds to constitutive principles, represented by ≫, and the third to epistemic principles, represented by >. Necessary principles entail both constitutive and epistemic principles; constitutive principles imply epistemic principles, but only defeasibly.

Since, on my account, constitutive and epistemic principles have essentially the same logic, one may wonder why the distinction matters. In ethics and other practical contexts, it is the difference between *prima facie* obligation and generalization about actual obligation. To use a different vocabulary, it is the difference between ends and means, between rules of action and rules of thumb, between something's being a primary reason for acting and a purely secondary reason for acting. (See, for example, Urmson 1975.) More broadly, it is the difference between generalizations that are lawlike and those that are distributional; it is the difference between what is true of something by virtue of what it is and what it true of it generally. These are distinctions with substantial contemporary significance and a rich philosophical history. Perhaps this paper has shed light on why they do not have a similarly rich logical history.

REFERENCES

Asher, N. (1995). Commonsense Entailment: A Conditional Logic for Some Generics. In G. Crocco, L. Fariñas del Cerro, and A. Herzog (Eds.), *Conditionals: From Philosophy to Computer Science.* Oxford: Clarendon Press. pp. 103–146.

Asher, N. and D. Bonevac (1996). Prima Facie Obligation. *Studia Logica* **57**, 19–45.

Asher, N. and D. Bonevac (1997). Common Sense Obligation. In D. Nute (Ed.), *Defeasible Deontic Logic.* Dordrecht: Kluwer. pp. 159–203.

Asher N. and M. Morreau (1991). Common Sense Entailment: A Modal Theory of Nonmonotonic Reasoning. *International Joint Conference on Artificial Intelligence* **91**, 387–392.

Asher, N. and M. Morreau (1995). What Some Generic Sentences Mean. In G. Carlson and J. Pelletier (Eds.), *The Generic Book.* Chicago: University of Chicago Press. pp. 300–338.

Benferhat, S., C. Cayrol, D. Dubois, J. Lang and H. Prade (1993). Inconsistency Management and Prioritized Syntax-based Entailment. *International Joint Conference on Artificial Intelligence* **93**, 640–645.

Bonevac, D. (1998). Against Conditional Obligation. *Noûs* **23**, 37–53.

Bonevac, D. (forthcoming). Predication *Per Se.*

Chellas, B. (1974). Conditional Obligation. In S. Stenlund (Ed.), *Logical Theory and Semantic Analysis.* Dordrecht: D. Reidel. pp. 23–33.

Chellas, B. (1980). *Modal Logic.* Cambridge: Cambridge University Press.

Dancy, J. (1991). An Ethic of Prima Facie Duties. In P. Singer (Ed.), *A Companion to Ethics.* Oxford: Blackwell. pp. 219–229.

Marcus, R. B. (1980). Moral Dilemmas and Consistency. *Journal of Philosophy* **77**, 121–136.

Morreau, M. (1997). Fainthearted Conditionals. *Journal of Philosophy* **94**, 187–211.

Urmson, J. O. (1975). A Defense of Intuitionism. *Proceedings of the Aristotelian Society* **75**, 111–119.

Williams, B. (1973). Ethical Consistency. In *Problems of the Self.* New York: Cambridge University Press. pp. 166–186.

Daniel Bonevac
Department of Philosophy
University of Texas at Austin
Austin, Tx 78712-1180
e-mail: dbonevac@mail.la.utexas.edu

Commentary by Mark Lance

For too long, philosophers of language focused their attention on strict, counterfactual, and indicative conditionals altogether ignoring the arguably more significant non-monotonic conditionals. That this situation is changing can only bode well for the level of clarity with which we address a range of issues in metaphysics and epistemology. Daniel Bonevac's "Constitutive and Epistemic Principles" further extends the scope of such reflection by urging, in a characteristically clear and impressive fashion, that there are two broad classes of non-monotonic conditional, one meta-physical and the other epistemic.[1]

This distinction between constitutive and epistemic principles is crucial, but perhaps only a starting point. That is, it seems to me that the epistemic principles Bonevac discusses are characterized only vaguely, while the notion of a constitutive principle is ambiguous. I'll not go into the vagueness of the epistemic notion here, except to say that the intuitive clarifications given in terms of 'good evidence for', 'on-necessary generalizations', 'defeasible warrant', etc. are neither altogether clear in the absence of an underlying epistemic theory, nor obviously equivalent.

In the case of constitutive principles, it seems to me that we clearly have three relations that might be expressed by the non-monotonic conditional form studied under Bonevac's heading "constitutive." $A \gg B$ could be read as: 'A is the (a) *cause of B*', 'A is partly *metaphysically constitutive of* (what it is to be) B (or vice versa?)', or 'B is true *because of A*'. (A clear example of the last notion would be 'If Joe cuts Sam, he will do something evil'. I take it that cutting is not even partly constitutive of evil, certainly that its role in the metaphysical nature of evil is insufficient to ground a *prima facie* inference of this sort. Nor is the point merely evidential, and there is certainly no sort of causal relation being expressed. What one wants to say is that cutting would, on this occasion, explain the fact that the act is evil, it is evil *because* it is a cutting.)

I would be very surprised if these notions turned out to be equivalent, and equally surprised if they lacked important non-identity relations between one another. Hopefully Bonevac's paper will inspire someone — perhaps even myself — to explore this and related issues.

With these brief remarks, I turn to the formal theory of constitutive and epistemic principles. Here, I'm less hopeful that Bonevac's paper will lead us in the right directions. I offer three sorts of counterexamples to various aspects of the formal theory, especially the account of "allowability" as an analysis of defeasible implication, and the truth conditions for epistemic conditionals.

Counterexample 1. Let Q be the claim that there are no irregularities in the world (where by 'irregularity' I mean a true instance of $\sim [(A > B) \longrightarrow (A \longrightarrow B)]$). Then consider a counterworld, w, to the inference from arbitrary P to Q. w must be a world with irregularities. Hence, if

---

[1]I have argued for a similar distinction, though drawn in rather different terms, in the monotonic realm. There are, I claim, at least two sorts of entailment, one having to do with commitment and the other with epistemic entitlement, and these are governed by different logics. In a larger work, I would attempt to relate those notions to Bonevac's. See Lance (1994).

there is any world w' with no irregularities, and P is true in w', then the inference from P to Q is allowed. But it is absurd to suppose that this is a reasonable defeasible implication.

Counterexample 1a. One might respond to this by denying that there are regular worlds.[2] One might also object to Counterexample 1 on the grounds that P is not the official language. This strikes me as a bit of a dodge, but all we need for the absurd result is some Q of the form $(A > B) \longrightarrow (A \longrightarrow B)$ and some sentence P such that for every world in which P is true Q false, there is a more regular world in which they are also. Then the inference from P to Q will be allowed.

Perhaps one could claim that arbitrary sentences *are* prima facie evidence for Q, on the grounds that $(A > B)$ is itself prima facie evidence for $(A \longrightarrow B)$. First, even if this latter epistemic connection holds, it hardly follows that arbitrary claims provide epistemic support for the claim that it does. Second, consider an instance we know to be false, i.e. in which $(A > B) \longrightarrow (A \longrightarrow B)$ is an irregularity. Say A is the claim that the referee declared the Dodgers to have won on Tuesday, and B the claim that the Dodgers did win. But suppose we know that in fact the referee was overruled by the league. Then it is clearly not true that 'Snow is white' is prima facie evidence for $(A > B) \longrightarrow (A \longrightarrow B)$.

Counterexample 2. Let Q be some necessary truth. Then there are no worlds which are counter-examples to the inference from P to Q so that inference is trivially allowed, for any P. Obviously not all such inferences are defeasibly reasonable. Nor is this because necessary truths fall outside the scope of defeasible reasoning. One can have good evidence for the truth of a logically necessary claim, or a necessary falsehood for that matter. (In my contribution to this volume, I explain in some detail, such a case.)

Counterexample 2a. Similar problems arise concerning the substitution of necessary identicals. Though 'the cardinality of the set of counterexamples to the well ordering axiom, given that the axiom of choice is true' is necessarily co-referential with 'the null set', we don't want to assume that (in all possible contexts) 'The null set is F' entails (or even defeasibly

---

[2]I for one have no intuitions at all about this matter or even about whether there are worlds which are more regular than a given world. Does w have *all* the irregularities of w'? Well that depends on which irregularities are true in each, and I have no intuitive sense of that.

implies) 'the cardinality of the set of counterexamples to the well ordering axiom, given that the axiom of choice is true is F'.

Counterexample 3. It is a central motivating claim of Bonevac's discussion that $A \gg B$ is defeasible justification for $A > B$. Now I don't quarrel with this inference per se – indeed, I find his discussion of this inference and its importance illuminating and convincing — but I deny that the inference is allowed, unless we accept implausible conceptions of when $A > B$ is true in a world. The key point is that $A > B$ can be false without $A \longrightarrow B$ being false. In his discussion of cases Bonevac often slides from a discussion of conditionals, to examples which are quantified conditionals. But even if we consider quantified principles, $(x)(Fx \longrightarrow Gx)$ can be true, without Fx being defeasible reason for Gx. (The point is obvious in the case of simple conditionals.) When he discusses the falsity of $(x)(Fx > Gx)$ Bonevac typically assumes that this must involve a number of case in which Fa and $\sim Ga$, but all that is necessary is that it be reasonable given our evidence, in many cases, to *think that Fx* and $\sim Gx$. Thus, consider the following claim:

> (1) Doctor x helps mother y deliver a child immediately after performing an autopsy without gloves or sterilization $\gg y$ develops an infection.

I take it that this claim has always been true in the actual world, when the connection is read causally. But if we replace the constitutive arrow with an epistemic one, and imagine the claim being made in 19th century Europe, it is false. Not only did the people lack the relevant background information to warrant this move, their background beliefs were such that the move would reasonably have seemed absurd. Hence, the actual world is one in which the causal version of (1) is true and the epistemic version false. But to specify the world this way is not to introduce *any* irregularities into the world.

What, then, are we to make of these counterexamples? Though these brief remarks are obviously not conclusive, I think there are two lessons to be learned. The idea of regularity is not a good one for capturing the truth of defeasibly reasonable inferences or epistemic conditionals. This is because the reasonability of these inferences has to do with what we can reasonably believe about A and B, not whether they are true, even across various worlds. Second, and related, possible worlds are not a likely candidate for a semantic treatment of these matters, for they are simply too blunt an instrument. Epistemic rationality can cut contents very finely. In fact, I am tempted to claim that if it is possible to believe

P and fail to believe Q, then there will be some situation in which both are rational.

To my mind, this all suggests we look to other tools for the explication of epistemic concepts. But even if I'm right in this, the work done by "Constitutive and Epistemic Principles" and other papers in the area by Bonevac is important, for it orients us to the right questions, makes many of the most important distinctions, and offers precise systematic analyzes to which to react. These are surely the things we most hope for in a philosophical analysis.

REFERENCES

Lance, M. (1994). Two Concepts of Entailment. *The Journal of Philosophical Research* **20**, 113–137.

Mark Lance
Department of Philosophy
Georgetown University
Washington, DC 20057-1076
e-mail: *lancem@guvax.acc.georgetown.edu*

Bonevac's Reply to Lance

Mark Lance raises several important issues in his response to my paper. Some pertain to my distinction between constitutive and epistemic principles. Others pertain to the Asher-Morreau theory of commonsense entailment I assume. I am grateful to him for the chance to clarify them.

1. The Constitutive/Epistemic Distinction

General principles, I claim, fall into at least two kinds. *Constitutive* principles assert that satisfaction of the subject term is *responsible* for satisfaction of the predicate term. *Epistemic* principles assert that satisfaction of the subject term is a *good indicator* of satisfaction of the predicate. The general idea is that the link between subject and predicate may be either metaphysical or epistemic. Lance is no doubt justified in complaining that this is vague, and that under each category are various distinct, nonequivalent notions. Satisfaction of the subject term may be responsible for or a good indicator of satisfaction of the predicate term in a number of ways, some of which may require adding constraints to my account or even changing the basic units of analysis, for example, from worlds to situations. (See, for example, the logic of causation in Koons 1998.) The existence of various subcategories of principles does not, however, vitiate my broader categorization.

Nevertheless, I mean to exclude a kind of epistemic connection that Lance emphasizes. I never speak of epistemic principles as being ones in which satisfaction of the subject term is defeasible warrant for satisfaction of the predicate term. (Indeed, the term *warrant* never appears in my paper.) Defeasible warrant must be modeled not by a principle or conditional but by defeasible implication. Quine has stressed that the material implication of the classical conditional must not be confused with entailment *per se*. Similarly, the epistemic conditional I have paraphrased in terms of indicating and providing evidence must not be confused with defeasible implication. The former is an object-language concept; the latter is metalinguistic. In commonsense entailment, the distinction is even more critical than usual, for there is no deduction theorem. Conditionalization is not a legitimate rule of inference.[3]

Lance's infection example illustrates a deeper reason for not thinking of epistemic principles in terms of defeasible warrant. What is defeasibly warranted in a given circumstance depends on the information available in that circumstance. But whether satisfaction of the subject term is a good indicator of satisfaction of the predicate term does not depend on available information.[4] Delivering a child after performing an autopsy without gloves or sterilization is a good indicator of later infection, and was so even before Semmelweis, even though no one knew it to be so.

We may use the distinction between epistemic principles and defeasible implication to clarify the point. Given only the information that the doctor delivered a child immediately after performing an autopsy, without

---

[3]In classical logic, $A, B \models A$. Since defeasible implication is supraclassical, A, B defeasibly imply A. Conditionalization would allow us to conclude that A defeasibly implies $B > A$. But that is plainly unacceptable: *You won't die today* does not defeasibly imply *If I cut off your head, you won't die today*. Moreover, if conditionalization were legitimate, A would defeasibly imply $\top > A$, i.e., $\square A$. On the analysis offered in "Constitutive and Epistemic Principles," then, constitutive and epistemic principles would be defeasibly equivalent.

[4]$A > B$ is true at a world w iff B is true at all A-normal worlds relative to w. There is no reference in this truth condition to anyone's state of information. $A > B$ is thus false at w iff B is false in some A-normal world relative to w. Since A-normal worlds are A-worlds, it follows that, if $A > B$ is false at w, there is some world w' which is A-normal relative to w where A is true and B is false. Lance is therefore right to say that $A > B$ can be false (in w) without $A \longrightarrow B$ being false (in w); if it were not so, $A \longrightarrow B$ would entail $A > B$. (So, therefore, would both B and $-A$.) But, if $A > B$ is false in w, $A \rightarrow B$ must be false in some w' A-normal with respect to w. In the quantified case, if $(x)(Fx > Gx)$ is false in w, there is an object a and a world w' Fa-normal with respect to w where $Fa \longrightarrow Ga$ is false.

gloves or sterilization (D), we cannot infer that the mother will develop an infection (I): D does not by itself defeasibly imply I. But $D > I$ is true; indeed, it is just the fact that Semmelweis noticed and then explained by discovering the causal connection that underlies the truth of $D \gg I$, which in turn explains $D > I$.[5] And $D > I$ and D together do defeasibly imply I.

I thus think Lance is wrong to conclude,

> The idea of regularity is not a good one for capturing the truth of defeasibly reasonable inferences or epistemic conditionals. This is because the reasonability of these inferences has to do with what we can reasonably believe about A and B, not whether they are true, even across various worlds.

The idea of regularity is supposed to capture only one of these notions, that of reasonable inference, which does indeed have to do with what can reasonably be believed. A defeasibly implies B if every counterexample — every world in which A is true and B false — is unnecessarily irregular. This definition seems to proceed strictly in terms of truth and falsehood, but the underlying idea is that it is reasonable to believe that things are regular in the absence of contrary information.

The truth of an epistemic conditional $A > B$, in contrast, depends on how things are in epistemically A-normal worlds. It does not depend on anyones state of information. It could be explicated in purely probabilistic terms without appeal to beliefs, except insofar as A and B themselves might appeal to beliefs.

---

[5]Lance's Counterexample 3 starts as a worry about whether inferences from constitutive to epistemic principles are allowed. Given the analysis of constitutive principles I eventually adopt, this should be clear. My earlier, informal motivation for the inference, however, which I offered before stating truth conditions even for epistemic principles, is unclear in a way that justifies Lance's worry:

To hold that arguments of the form

    (15) $\forall x (Fx \gg Gx)$
         $\therefore$ $\forall x (Fx > Gx)$

are allowed is to maintain that all counterexamples to such arguments are unnecessarily irregular, that is, that all counterexamples involve modus ponens failures not required by the premise. Suppose that the conclusion of (15) is false. Then, intuitively, there must be some normal (typical, standard) objects a such that Fa but not Ga. The premise of (15), however, implies that $Fa \gg Ga$. Any counterexample to (15), then, is a world in which, for some normal a,

    (16) $(Fa \gg Ga) \longrightarrow (Fa \longrightarrow Ga)$

is false. But that means that any counterexample to (15) is constitutively irregular.

This is sloppy, for $Fa \longrightarrow Ga$ may fail in a world other than the one in question. To get the explanation to work, one would have to know that if a constitutive conditional $Fa \gg Ga$ is true in w, it is also true in the Fa-normal worlds relative to w. This is what my latter analysis of $A \gg B$ as $T > (A > B)$ supplies.

2. Commonsense Entailment

Lance's chief objection to my account of epistemic conditionals thus leads to the heart of the Asher-Morreau theory, the distinction between faint-hearted conditionals and defeasible implication. Theorists inspired by Sellars-Brandom, Gauker, Lance, and O'Leary-Hawthorne, for example, try to analyze inference in terms of assertability, without appealing to such a distinction. Whether such a project can succeed is beyond the scope of these remarks. A related issue, however, underlies Lance's other objections, aimed directly at commonsense entailment.

Counterexample 1. Lance observes that in commonsense entailment, P defeasibly implies $(A > B) \longrightarrow (A \longrightarrow B)$, so long as P is independent of A and B.[6] He takes this to be a bug; I consider it a feature. Assume that S defeasibly implies $A \longrightarrow B$ iff $S \cup \{A\}$ defeasibly implies B. (Assume, in other words, that modus ponens and conditionalization hold for the material conditional with respect to defeasible implication. This is true in commonsense entailment and various other systems of nonmonotonic reasoning, e.g., circumscription.) Then P defeasibly implies $(A > B) \longrightarrow (A \longrightarrow B)$ iff $\{P, A, A > B\}$ defeasibly implies B. Where P is independent of A and B, this is just the principle of irrelevant information, which surely ought to hold. If Tweety is a bird, and birds fly, it is reasonable to conclude that Tweety flies – even given the additional information that it is Tuesday, that Sylvester climbs trees, or that Granny is knitting.

Lance's natural language counterexample cheats in a revealing way:

> Say A is the claim that the referee declared the Dodgers to have won on Tues-
> day, and B the claim that the Dodgers did win. But suppose we know in fact
> that the referee was overruled by the league. Then it is clearly not true that
> 'Snow is white' is prima facie evidence for $(A > B) \longrightarrow (A \longrightarrow B)$.

Lance's example shows only that the inference from 'Snow is white' to $(A > B) \longrightarrow (A \longrightarrow B)$ is defeasible, not that it is disallowed. The kind of implication at stake here is defeasible, so any information available must be taken into account explicitly. The information that the referee (or official scorer, if the sport is baseball; baseball has no referees) was overruled by the league is not independent of A and B; in context, it implies $A \& -B$. Representing the example fairly, then, we can ask whether $\{$'Snow is white', A, $-B\}$ defeasibly implies $(A > B) \longrightarrow (A \longrightarrow B)$. Commonsense entailment provides the intuitively correct answer: No.

---

[6]This is actually Lance's Counterexample 1a; Counterexample 1 cannot be stated without propositional quantification, which introduces considerable and irrelevant complications.

Counterexample 2. Anything defeasibly implies a necessary truth, Lance observes. Similarly, contradictions defeasibly imply anything. And substitution of necessary identicals preserves defeasible implication. All these features of commonsense entailment follow from the supraclassicality of defeasible implication. It might be better, since the notion is supraclassical, to refer to it more carefully as (possibly) defeasible implication. As Lance's mathematical examples show, the issue is independent of defeasibility.

I can see three options short of moving to an assertability theory: (a) to bite the bullet, using Gricean considerations to dispel the air of peculiarity around some of these inferences and reminding the reader that what is expressed by constitutive and epistemic conditionals is independent of states of information. (b) to weaken the background modal logic, adding an accessibility relation to the semantics. In the absence of constraints linking accessibility to normality, this eliminates the questionable inferences. In effect, it allows one to contemplate epistemically normal but nevertheless impossible worlds. (c) to discriminate contents more finely, as one must do in any case to account for the logic of belief and other attitudes.

Like Lance, I am inclined to see option (c) as the most promising. Unlike him, however, I envision commonsense entailment meshing nicely with a theory of fine-grained contents. In fact, earlier and technically more cumbersome presentations of commonsense entailment used information states explicitly in defining defeasible implication (Asher and Morreau 1995). The possible-worlds analysis of information states, implicit in the Morreau (1997) presentation on which I rely, yields easily to an analysis in terms of situations (Koons 1998) or representational structures such as DRSs. (See Kamp 1981; Asher and Lascarides 1993; Asher 1995.)

Even so, the coarser analysis I have offered may suffice in many contexts, even ones where representational states are involved. Does Kripke's Pierre, for example, have information from which it would be reasonable to conclude that Londres is not pretty? If there is a sense in which the answer is yes — as I think there is — then a coarse-grained analysis not only marks a step to a more adequate analysis but captures a viable notion of defeasible inference in its own right.

REFERENCES

Asher, N. (1995). *Reference to Abstract Objects in Discourse.* Dordrecht: Kluwer.

Asher, N. and A. Lascarides (1993). Temporal Interpretation, Discourse Relations and Commonsense Entailment. *Linguistics and Philosophy* **16**, 437–493.

Asher, N. and M. Morreau (1995). What Some Generic Sentences Mean. In G. Carlson and J. Pelletier (Eds.), *The Generic Book.* Chicago: University of Chicago Press. pp. 300–338.

Morreau, M. (1997). Fainthearted Conditionals. *Journal of Philosophy* **94**, 187–211.
Koons, R. (1998). Realism: Applications of an Exact Theory of Causation and
    Teleology. *Unpublished Manuscript.*

Poznań Studies in the Philosophy
of the Sciences and the Humanities
2000, *vol.* 71, *pp.* 219–242

Otávio Bueno

EMPIRICISM, MATHEMATICAL TRUTH AND MATHEMATICAL KNOWLEDGE [1]

ABSTRACT. According to Benacerraf, the current interpretations of mathematics face a dilemma when they try to present simultaneously appropriate notions of mathematical truth and of mathematical knowledge: a suitable characterization of one implies the inadequacy of the other. In this paper, I outline a constructive empiricist interpretation of mathematics, according to which these two notions can be held together. The main strategy consists in countenancing a weaker notion of truth and a broader concept of structure (da Costa's and French's quasi-truth and partial structure), and in terms of them, to develop a correspondingly broader approach to mathematical knowledge.

1. Introduction

In an influential paper, Paul Benacerraf has argued that the current interpretations of mathematics face a dilemma when they try to present simultaneously appropriate notions of mathematical truth and of mathematical knowledge (see Benacerraf 1973). In his view, a suitable characterization of one implies the inadequacy of the other: an adequate notion of mathematical truth leads to an inadequate concept of mathematical knowledge, and an appropriate notion of mathematical knowledge leads to an inappropriate concept of mathematical truth (see also Hellman 1989, p. 1).

In this paper, I am concerned with outlining an empiricist interpretation of mathematics, according to which suitable notions of mathematical truth and mathematical knowledge can be held together. The main strategy consists in countenancing a weaker notion of truth and a broader concept of structure — namely, quasi-truth and partial structure, as developed by Newton da Costa and Steven French in a series of works — and in terms of them, to develop a correspondingly broader approach to

---
[1] Many thanks to Steven French, Newton da Costa and James Ladyman for helpful discussions on the issues examined here.

knowledge.[2] As a result, or so I shall argue, both mathematical truth and mathematical knowledge can be accommodated. It should be stressed that here I am only concerned with sketching the general formal framework, and I shall leave the details for another opportunity.

Further constraints come from the empiricist side of this account. The main idea is to extend to mathematics the best articulated empiricist interpretation of science. In my view, Bas van Fraassen's constructive empiricism supplies such an interpretation (see van Fraassen 1980, 1985, 1989, 1991). One of the chief aspects of constructive empiricism is that, in order to interpret science, one does not need to be committed to a particular and problematic kind of metaphysics, called by van Fraassen *postulational metaphysics*, which countenances the introduction of non-observational entities in the explanation of "phenomena."[3] Bluntly put, postulational strategies are problematic for the empiricist since they are devised in order to legitimate the introduction of non-empirical factors in knowledge acquisition. That constructive empiricism allows non-empirical factors in the interpretation of science is perhaps one of its most salient features — after centuries of empiricist accounts which can be regarded as too restrictive. Consider, for instance, the role played by these factors in the interpretations of quantum mechanics. Each interpretation provides an understanding of how the world could be, and to this extent, given the concern for informativeness, the empiricist is engaged in the interpretative process (see van Fraassen 1991). According to van Fraassen, in some cases an interpretation may even lead to the formulation of a new theory, and this is an extreme case in which the development of interpretations can be accommodated on empiricist grounds (see van Fraassen 1989, pp. 226–227).

Nevertheless, an interpretation of a scientific theory is not a piece of postulational metaphysics, given that no entities are introduced to provide explanations for the "phenomena." All we have is the free exploration of *possibilities* — namely, of how the world might be if the theory under interpretation were true (see van Fraassen 1989, 1991). Likewise, in my view, the interpretation of a mathematical theory within a constructive empiricist framework also brings a *modal* component: one should examine what mathematical statements hold (or, as I shall argue, are quasi-true) if the convenient structures were possible. (As we shall see in Section 4, there is a close relationship between this proposal and Hellman's modal-

---

[2]For further details about the concepts of quasi-truth and partial structures, which constitute the partial structures approach, see (da Costa and French 1989, 1990, 1993a, 1993b, 1995, forthcoming).

[3]A critical discussion of this metaphysics can be found in (van Fraassen 1993b, 1994).

structural interpretation; see Hellman 1989.)

However, why should anyone care about an empiricist interpretation of mathematics? For the empiricist, such an interpretation extends empiricism to a new domain, and thus provides evidence for the program as a whole. For the non-empiricist, the account supplies an interpretation which is not tied to a problematic "postulational metaphysics," avoiding in this way the postulation of non-observational entities. (I take it that even those who have nothing against these entities, for ontological parsimony, if nothing else, would prefer a view which does not depend on them.) In particular, within mathematics, the empiricist advances an interpretation which is not committed to abstract objects (such as sets, functions and so on). And if it is possible to travel with a metaphysically lighter luggage, why not do so?

Two main components are then articulated: the first is to adopt a convenient version of *structuralism* (which supplies a rich framework to represent the subject-matter of mathematics); the second is to combine this structuralist component with a formulation of *modalism* (which provides a strategy to avoid the commitment to abstract objects). The first component stresses the structural features of mathematics, leaving mathematical objects behind; the second is meant to evade an ontological commitment even to structures. Taken together, we have a proposal in which both the structural and the modal components of constructive empiricist are accommodated (see Bueno 1997b).

This paper has three further sections. Section 2 is concerned with the presentation of Benacerraf's dilemma, and will set the scene for the arguments I will present in the sequence. In Section 3, I shall spell out the formal framework with which I will be working — the partial structures approach — emphasizing, in particular, the notions of partial structures and quasi-truth. Finally, in Section 4, I will argue that, in terms of this framework, an empiricist account of mathematical truth and mathematical knowledge can be sketched, circumventing in this way the challenge put forward by Benacerraf.

2. Benacerraf's Challenge

In his 1973 paper, Benacerraf puts forward two adequacy conditions for any philosophical account of mathematical truth and mathematical knowledge, and argues that none of the extant accounts meet them. The first condition is that any minimally adequate notion of mathematical truth should imply that truth conditions for mathematical statements are clearly conditions for their truth. In his view, what is required is an "overall theory of truth in terms of which it can be certified that the account of mathematical truth is indeed an account of mathematical *truth*" (Benacerraf 1973, p. 18). Otherwise, the putative characterization of mathematical truth is not going to be substantial enough. The only account of truth which can be used as the basis for a notion of mathematical truth, according to Benacerraf, is Tarski's (as developed in Tarski 1983), and thus "any putative analysis of mathematical truth must be an analysis of a concept which is a truth concept in at least Tarski's sense" (Benacerraf 1973, p. 19). It is crucial for Benacerraf's argument to claim that an "essential feature [of Tarski's account] is to define truth in terms of reference (or satisfaction) on the basis of a particular syntactico-semantical analysis of the language" (p. 19). Since, according to Benacerraf, Tarski's account is committed to *reference*, a substantial notion of truth is provided, given that a domain of objects is assumed and truth is a feature of the sentences of the language under consideration, involving "the reference of the singular terms and predicates" (p. 19). I call this the *semantic* requirement.

The second condition is concerned with mathematical knowledge. In Benacerraf's view, "an account of mathematical truth, to be acceptable, must be consistent with the possibility of having mathematical knowledge: the conditions of the truth of mathematical propositions cannot make it impossible for us to know that they are satisfied" (p. 19). More strongly, he insists, "the concept of mathematical truth . . . must fit into an over-all account of knowledge in a way that makes it intelligible how we have the mathematical knowledge that we have" (p. 19). I call this the *epistemological* requirement.

According to Benacerraf, these two conditions rule out all the extant accounts of mathematical truth. And this is because the proposals more sensitive to the epistemological requirement — which avoid the commitment to mathematical entities — do not supply an adequate characterization of mathematical truth. Moreover, the proposals which appropriately meet the semantic requirement — and assume the existence of mathematical objects — cannot explain how our knowledge of these causally inert objects is possible.

This is an ingenious argument and it is by no means surprising that it has generated so much discussion. If so much of it seems to be inconclusive, it is perhaps because of the lack of an appropriate framework. In what follows, in terms of da Costa's and French's partial structures approach, I shall propose a fresh start.

3. Partial Structures and Quasi-Truth

In a seminal paper, Mikenberg, da Costa and Chuaqui presented a conceptual framework articulated in terms of partial structures and quasi-truth (see their 1986). This framework has been later extended, by da Costa and French, to accommodate several issues in the philosophy of science, including the interpretation of probability theory (see da Costa 1986), the logic of induction (see da Costa and French 1989), the model-theoretic approach (see da Costa and French 1990), the theory of acceptance of theories (see da Costa and French 1993a), as well as the modeling of "natural reasoning" (see da Costa and French 1993b). I shall spell out now the main tools provided by this proposal.

The investigation of a particular domain M of knowledge generally involves the study of certain relations among its objects. However, the information about these objects is in general "incomplete," in the sense that it is not known whether the relations concerned are applicable to every (n-tuple of) object(s) of the domain under consideration. In order to formally accommodate this situation, the concept of a *partial relation* has been introduced. According to da Costa and French, a relation is *partial* in the sense that it is not defined for every object (or n-tuple of objects) of the domain D in question. More formally, an n-place partial relation R is a triple $\langle R_1, R_2, R_3 \rangle$, where R_1, R_2, and R_3 are mutually disjoint sets, with $R_1 \cup R_2 \cup R_3 = D^n$, and such that R_1 is the set of n-tuples that satisfy R; R_2 the set of n-tuples that do not satisfy R; and finally R_3 of those n-tuples for which it is not defined whether they satisfy R or not. (It should be noticed that when R_3 is empty, R is a standard n-place relation that can be identified with R_1; see da Costa and French 1990, p. 255, note 2.)

However, the representation of our patterns of modeling information requires more than partial relations: an adequate notion of *structure* should also be presented. This notion is likewise conceived as encompassing the "openness" typical of our epistemic situation where we often deal with "incomplete" information. Its use is also meant to stress the particular role played by structures in the process of representing bits of empirical information. Meeting these demands, and based on this notion of a partial relation, it is natural to introduce the concept of a *partial structure*. A *partial structure* is an ordered pair $\langle D, R_i \rangle_{i \in I}$, where D is a non-empty set (representing the objects employed in the systematization of the relevant domain of knowledge M, whose study we are concerned with), and $(R_i)_{i \in I}$ is a family of partial relations, in the sense just presented, defined over D.

These structures can be used as basic components to model certain aspects of scientific activity, in particular in those cases in which models are employed in certain branches of science (see da Costa and French 1990, forthcoming). But they also have a second, more "formal," function. They can be used in the formulation of a particular notion of truth, which extends Tarski's account and leads to the characterization of the concept of *quasi-truth*. Partial structures display here nearly the same role that the formal concept of interpretation (which is formulated in terms of full structures) has in the Tarskian semantics: if truth is defined in terms of an interpretation, quasi-truth is defined in terms of a partial structure.

The connections between truth and quasi-truth are still tighter. The strategy of defining the latter consists in introducing an "intermediary" kind of structure so that the former can be employed. Given that Tarskian semantics was constructed only for full structures, it is necessary that a total structure be obtained from a partial one by a process of "filling in" its partial relations. We call these "filled in" structures *normal structures*. More formally, given a partial structure $A = \langle D, R_i \rangle_{i \in I}$, we say that the structure $B = \langle D', R_i' \rangle_{i \in I}$ is an *A-normal structure* if the following conditions are satisfied: (1) $D = D'$; (2) every constant of the language under consideration is interpreted by the same object both in A and in B; and (3) R_i' extends the corresponding relation R_i (in the sense that, as opposed to the latter, the former is defined for every n-tuples of objects of its domain). It should be noticed that given a partial structure, it is not always the case that we can extend it into a normal one. Necessary and sufficient conditions for this result are presented in Mikenberg, da Costa and Chuaqui (1986) (for a discussion, see da Costa, Bueno and French 1998; Bueno 1997a, Section 3.1; Bueno and de Souza 1996).

It is clear that A-normal structures are constructed in order to provide an interpretation of the language under consideration. This was, to a certain extent, the strategy devised by Tarski to formulate rigorously the concept of truth: the latter is defined in a *structure*. The same feature is also found in the formulation of quasi-truth. We say that a sentence E is *quasi-true* in a partial structure $A = \langle D, R_i \rangle_{i \in I}$, according to the A-normal structure $B = \langle D', R_i' \rangle_{i \in I}$, if E is true in B (in the Tarskian sense). If E is not quasi-true in S according to B, we say that E is quasi-false (in S according to B).

In my view, this framework supplies important resources to examine Benacerraf's dilemma (from an empiricist view), given the broader notion of structure it provides and the "openness" of its conceptual features. It is now time to substantiate this claim.

4. Mathematical Truth and Mathematical Knowledge

Two steps should be taken by the empiricist in order to circumvent Benacerraf's challenge. The first is to defend the idea that an appropriate notion of mathematical truth is supplied by quasi-truth. In this way, the first horn of Benacerraf's dilemma — the semantic requirement — is approached directly. According to the empiricist view I am suggesting here, a mathematical theory does not need to be true to be good, but only quasi-true (see also Bueno 1997b). This is of course something to be expected in a nominalist account of mathematics. Hartry Field, in the development of his nominalist program, has argued that truth is not something we should necessarily require of a mathematical theory; a different feature is to be highlighted instead: conservativeness (see Field 1980, 1989). The latter, Field stresses, is not weaker than truth, it is only a different aim (see Field 1982, p. 59; and Field 1988, pp. 240–242). The problem for the empiricist is that, as opposed to Field's proposal, a notion which is not only different, but also *weaker than truth* should be countenanced as the aim of mathematics — just as, in science, constructive empiricism has been articulated in terms of the claim that the aim of science is not truth, but something weaker, empirical adequacy (see van Fraassen 1980, p. 12; van Fraassen 1989, pp. 189–193). Since quasi-truth is weaker than truth, in this respect it is appropriate for an empiricist account.

But why does quasi-truth provide, at least in principle, a notion of mathematical truth? This is, of course, a delicate question, and it would be unreasonable to pretend that any simple answer would suffice. Despite this, let me outline some considerations. One of the virtues of quasi-truth for the articulation of a philosophy of mathematics comes from the fact that it is formulated in terms of Tarski's account (see Tarski 1983). As we saw, a sentence E is quasi-true in a partial structure A if there is an A-normal structure B in which E is true (in the Tarskian sense). However, Tarski's formulation of truth provides a formal framework in which *reference* to mathematical objects is supposed — and as we saw, this is a basic assumption in Benacerraf's dilemma. The idea is that truth, and thus satisfaction, is analyzed in terms of a domain of objects (mathematical objects), and so mathematical truth, if taken as *truth* in the Tarskian sense, leads to the commitment to mathematical objects. Clearly, this brings problems for empiricism, given its rejection of postulational metaphysics. However, by taking the Tarskian approach as basic, this concept of mathematical truth is at least appropriate to the extent that a notion of mathematical *content* (not open to a purely syntactic account of truth) can be accommodated: there is something mathematics is about, a domain of objects (which are not taken as syntactic entities), which is "grasped" by mathematical truth. The price that we apparently have to pay is Platonism: mathematics, in this construal, is about a world of non-empirical objects (and this is, of course, one of the horns of Benacerraf's dilemma). The problem then is how to reconcile this feature with empiricism.

It is precisely at this point that quasi-truth enters. This notion offers an account of mathematical truth in which something substantive can be said about truth (given the use of Tarski's characterization), without an ontological commitment to mathematical objects. There are two main strategies to be developed by the empiricist at this point. The first is that, by focusing on a partial structure A, the ontological commitment is restricted to those objects whose properties are defined in A (as we saw, each partial relation R has a R_3-component involving those n-tuples for which it is *not* defined whether they belong or not to R). This is already a way of avoiding certain ontological commitments. The second strategy, which is derived from the first, concerns the specific use of quasi-truth. Given the formal features of the latter, its use supplies a deflationary account of mathematical objects. If the aim of mathematics is quasi-truth, a mathematical theory does not need to map, in complete detail, every aspect of the domain to which it is applied. It suffices if it accommodates certain aspects of this domain. This point is perhaps clearer in connection to science. In the explanation of certain phenomena, we may introduce certain theoretical components which may not have a counterpart in reality — they are taken to be only features of the models we use. This is, of course, a cherished empiricist idea (see, for instance, van Fraassen 1989, pp. 92–93), and quasi-truth straightforwardly accommodates it: the quasi-truth of a theory T depends on the extension of the partial structures employed to model certain phenomena, and there is no claim that T is in complete agreement with every aspect of the domain under consideration (since there are different A-normal structures that extend a partial structure A to a full structure, a quasi-true theory is not necessarily true).

If we try to extend this picture to mathematics, the first problematic issue is to identify the *domain* of a mathematical theory. The reference to a domain may assume that we have from the outset a class of mathematical objects which constitute the subject of mathematical research. And if this is so, it would be pointless for the empiricist to try not to be committed to these objects. However, things are not so straightforward. The first point to be noticed is that, as opposed to Benacerraf's view, Tarski's characterization of truth is *not* committed to the notion of reference (let alone a causal theory of reference). All that is required is a notion of function, in terms of which the interpretation of a formal language is articulated (see Tait 1986, pp. 148–150). This function relates two mathematical structures: one results from the syntax of the language under interpretation, the other is the structure in which the language is interpreted. Nevertheless, the existence of this function is neither a necessary, nor a sufficient condition for reference. There are expressions in a language which certainly refer, but which are not functions. Consider, for instance, "the son of Mr. and Mrs. Wittgenstein," in the domain of human beings (the parents of Wittgenstein had more than two sons). Moreover, there are functions which do not refer. Take, for example, the 0-place function "the most famous detective in Conan Doyle's histories," in the domain of human beings. Thus, since all that is demanded in the Tarskian account is a function, and the existence of it does not guarantee *reference*, there is no commitment to the notion of reference in this account. And so, given that quasi-truth was defined in terms of the Tarskian proposal, the use of the latter in the characterization of the former does not lead *ipso facto* to Platonism. The talk of a domain of a mathematical theory — which is required in the application of the Tarskian account, and of the notion of quasi-truth, to characterize mathematical truth — does not commit the empiricist to the existence of mathematical objects.

A further putative difficulty with the use of the Tarskian view in mathematics comes from the fact that it is articulated in terms of a mathematical language (a particular set theory), and so we are interpreting mathematics within mathematics itself (this point is also made by Tait 1986, p. 149). However, given that the constructive empiricist has rejected a foundationalist view in science, ruling out the idea that there is a ultimate foundation of knowledge from which all knowledge claims have to be derived (see van Fraassen 1993a), the same rejection should be extended to the philosophy of mathematics. Thus, the regress involved in interpreting a mathematical theory within mathematics is not threatening. To be more specific, the role of an interpretation of mathematics is not to provide a *foundation* — in the foundationalist sense — to this field, but to supply certain resources to evaluate the acceptability of mathematical theories.[4] Such an acceptability is of course open to revision, and the epistemic issues brought by the acceptance of a mathematical theory can be accommodated in terms of the degrees of quasi-truth (see Bueno, forthcoming). Mathematical theories are *accepted* by the empiricist, given their quasi-truth, but he or she does not necessarily *believe* in them, since such theories can be false, due to the distinct ways of extending a partial structure to a full one. In this way, van Fraassen's well known distinction between belief and acceptance is retained in this construal (see van Fraassen 1980, 1989).

Granted that the empiricist may not believe in mathematical theories, does he or she have anything to say about *mathematical objects*? In other words, what are mathematical objects within the empiricist picture? With this question we reach the second point of this empiricist proposal,

---

[4]An ingenious defense of a non-foundationalist view in the philosophy of mathematics, including certain points that can be adapted by an empiricist view, can be found in Shapiro (1991).

which brings the *structuralist* component. Here is a slogan to describe
how mathematical objects are taken in this construal: they are *relations
among relations*. This idea, as we shall see, is a variation of some proposals advanced by structuralists in the philosophy of mathematics (see, for
instance, Resnik 1981; Shapiro 1989, 1991; Hellman 1989, 1996; Benacerraf 1965), as well as of structural realists in the philosophy of science
(see especially French, forthcoming and Ladyman 1998). However, due
to the crucial role of structures, on the one hand, and the lack of commitment to mathematical objects, on the other, it seems appropriate to
call this proposal *structural empiricism* in mathematics (see also Bueno
1997b, forthcoming).

Structural realism is an approach to science which constitutes an alternative to entity-realism — which countenances a realist interpretation
of "theoretical" *entities* — in which the epistemic role is played by structures, such as the mathematical structures employed in the formulation of
scientific theories. The basic idea is that, instead of entities or objects, all
we have to be realist about are structures. In his elaboration of this view,
French has argued that objects have only a heuristic role to lift us up to
the structures, and thus can be dispensed with in structural realism (see
French, forthcoming). I am quite sympathetic to this idea. My only worry
is about the notion of structure assumed in this move. The standard view,
in a simplified version, countenances that a structure is an ordered pair
$\langle D, R_i \rangle_{i \in I}$, where D is a set of *objects* and $(R_i)_{i \in I}$ is a family of relations
defined over D. However, in this characterization of structure, it is not
at all obvious how such objects can be dispensed with without damaging
the whole structure as it were, since they are constitutive elements of the
latter. What is needed is a new notion of structure, in which the objects
are "individuated," as it were, through the relations holding between the
relations of the structure — so that objects are a sort of "second-order"
derivative component of the structure. For instance, in mathematics, we
do not have to start with a set of numbers (such as D) and study the relations R_i between numbers. All we have are the relations, and we "regain"
the numbers (if we want) in terms of the relations among the relations
we already have. In arithmetic, for example, the number 4 is the greatest
number which is smaller than 5. What about 5? 5 is the greatest number which is smaller than 6, and so on. (This is how the infinity of the
natural numbers can be accommodated by their own "characterization.")
Of course, 0 is the number which is smaller than every number. Notice
that we are quantifying only over relations (smaller than, bigger than)
and numbers (which, in the present account, are viewed as relations between relations), and so *there is no quantification over objects*. Therefore,
objects lack any ontological priority, and can be dispensed with in the

structure.[5]

This proposal is, of course, connected to structuralist views in the philosophy of mathematics. According to these views, an abstract notion of structure should be countenanced in order to accommodate the various aspects of mathematics. Despite the differences between Resnik's, Shapiro's and Hellman's structuralist proposals, on at least one point all of them agree: a set-theoretical notion of structure is too particular to be taken as the basis for structuralism, since it is only one among several extant concepts of structure in mathematics. Resnik proposes instead the notion of *pattern* (see Resnik 1981). However, when this notion is spelled out, it is nothing but a restatement of the concept of structure found in set theory (with a domain D and relations defined over D). Shapiro (1989) talks of *systems of objects*, and explicitly acknowledges that, in order to accommodate set theory, which after all is a mathematical theory among others, his proposal has to be (and in fact is) a notational variant of set theory (see Shapiro 1989, p. 164).

Whereas both Resnik and Shapiro countenance Platonist views (according to which mathematics is about a domain of existing abstract structures), Hellman, advocating his modal-structural interpretation, introduces a nominalist version of structuralism. Roughly speaking, his idea — which develops in detail a proposal suggested in Putnam (1967) — is that mathematics is concerned only with *possible* structures, which are not necessarily actual, and so no commitment to mathematical objects is required in developing an interpretation of mathematics (for details, see Hellman 1989, 1996). In order to develop this idea, Hellman adopts a primitive notion of modality based on S5 (see Hellman 1989, p. 17). So instead of claiming that a certain arithmetical statement S is true, it suffices to show that it would hold in a structure of a convenient type, if there were such a structure (this is, roughly speaking, the *hypothetical* component). Moreover, the structures under consideration are *possible*, according to the primitive notion of modality (and this is the *categorical* component). Together the two components allow a nominalist treatment of mathematics, avoiding the ontological commitment to mathematical entities.

The main differences between the present empiricist proposal and the structuralist accounts are two-fold. (1) As opposed to Resnik's and

---

[5]It is worth noticing that this suggestion is not necessarily incompatible with French's point to the effect that objects are only convenient "tags" to lift us up to the structures. The idea is that, in mathematics, as opposed perhaps to physics, a more abstract notion of structure can be employed in order to avoid the commitment to (mathematical) objects. In physics, structural realists explore instead the role of invariances and symmetries in the "constitution" of physical objects (see Ladyman 1998 and French, forthcoming).

Shapiro's views, Platonism is avoided by articulating a modal proposal
similar to Hellman's. (2) However, as opposed to Hellman's conception,
modalities are not taken as primitive, but are interpreted in terms of
quasi-truth (see Bueno 1997b). The idea is to interpret mathematical
statements as modal statements (about the possibility of certain structures), and to claim that the aim of modal talk is only quasi-truth, and
not truth. (Given that, in this construal, mathematical discourse is a
particular case of modal discourse, the aim of mathematics is the same as
the aim of modal talk.) Modal operators of necessity and possibility can
then be analyzed in the following way:

Necessarily P is quasi-true (in a partial structure A) iff for all A-normal
structures, P is quasi-true.

Possibly P is quasi-true (in a partial structure A) iff there is some A-
normal structure in which P is quasi-true.

But what do we gain with such a deflationary move? In my view, in
this way not only can one of the chief objections to modalism be answered,
but also the second horn of Benacerraf's challenge can be accommodated.
These two points shall be considered in turn. The objection to modalism
was made by Shapiro (see his 1993, p. 466). In his view, it is unclear
why the introduction of modalities is of any help in the epistemology
of mathematics, given that instead of focusing on abstract objects, we
have to accommodate *possible* abstract objects — and no reason has been
advanced to the effect that the latter are more tractable than the former.
What is required here, I take it, is an appropriate modal epistemology. As
for Benacerraf's challenge, the issue concerns the possibility of knowledge
of mathematical objects, given that we have no causal interaction with
them — this is the epistemological requirement.

The main idea is to articulate a deflationary account of mathematics:
mathematical statements are understood as modal statements. As a result, no commitment to the existence of mathematical objects is required
(all we have are *possible* structures), and thus our knowledge of these objects is similar to our knowledge of other fictitious objects. It is derived
from the "properties" that these objects presumably have. Such "properties," especially in the case of mathematics, result from the structures
in which these objects are formulated — the objects themselves being
nothing but "relations among relations." Thus, instead of articulating
a modal epistemology, given that mathematical theories are taken to be
quasi-true (and not necessarily true), what is required is an epistemology
of structures.

The second step open to the empiricist to circumvent Benacerraf's challenge enters at this stage. Given the close connection between quasi-truth, scientific practice and scientific knowledge (spelled out, for instance, in da Costa and French 1990, forthcoming; French 1997 and Bueno 1997a), the knowledge horn of the dilemma is considered by articulating the structural components found in scientific and mathematical practice. Mathematical knowledge based on quasi-true theories is the result of the construction of certain partial structures and the study of their extension to full ones. The pluralism of different mathematical theories is accommodated to the extent that there is a core of "accepted structures" whose extensions differ. It is a sort of "quasi-knowledge," as far as partial structures are concerned, which increases with the increase of information about the domain under consideration.

Moreover, as for the lack of causal interaction with mathematical objects, we should recall that the constructive empiricist has rejected causality even at the empirical level — the latter is, at best, a feature of our models (see van Fraassen 1980, 1989). Therefore, it is question-begging to assume a causal account of knowledge, which is a basic premise in Benacerraf's challenge, in an argument against empiricism in mathematics. If mathematical objects are taken as convenient fictions, it is by no means surprising that they are causally inert.

In this way, by suitably changing the notion of truth, and countenancing quasi-truth as an aim of mathematics, the empiricist can address the two horns of Benacerraf's dilemma. Despite the fact that several points have still to be developed, we have here at least an outline of a constructive empiricist epistemology of mathematics.

REFERENCES

Benacerraf, P. (1965). What Numbers Could Not Be. *Philosophical Review* **74**, 47–73. (Reprinted in Benacerraf and Putnam (Eds.), 1983, pp. 272–294.)

Benacerraf, P. (1973). Mathematical Truth. *Journal of Philosophy* **70**, 661–679. (Reprinted in Hart 1996, pp. 14–30. Page references in the text are to this reprint.)

Benacerraf, P., and H. Putnam (Eds.), (1983). *Philosophy of Mathematics: Selected Readings* Second edition. Cambridge: Cambridge University Press.

Bueno, O. (1997a). Empirical Adequacy: A Partial Structures Approach. *Studies in History and Philosophy of Science* **28**, 585–610.

Bueno, O. (1997b). Structure, Modality and Quasi-Truth. Paper read at the History and Philosophy of Science Seminar, Department of Philosophy, University of Leeds.

Bueno, O. (forthcoming). What is Structural Empiricism? Scientific Change in an Empiricist Setting. Forthcoming in *Erkenntnis*.

Bueno, O., and E. de Souza (1996). The Concept of Quasi-Truth. *Logique et Analyse* **153/154**, 183–199.

Butterfield, J., and C. Pagonis (Eds.), (forthcoming). *From Physics to Philosophy*. Cambridge: Cambridge University Press.

Churchland, P.M., and C.A. Hooker (Eds), (1985). *Images of Science: Essays on Realism and Empiricism.* Chicago: The University of Chicago Press.

da Costa, N.C.A. (1986). Pragmatic Probability. *Erkenntnis* **25**, 141–162.

da Costa, N.C.A., O. Bueno, and S. French (1998). The Logic of Pragmatic Truth. *Journal of Philosophical Logic* **27**, 603–620.

da Costa, N.C.A. and S. French (1989). Pragmatic Truth and the Logic of Induction. *British Journal for the Philosophy of Science* **40**, 333–356.

da Costa, N.C.A. and S. French (1990). The Model-Theoretic Approach in the Philosophy of Science. *Philosophy of Science* **57**, 248–265.

da Costa, N.C.A. and S. French (1993a). Towards an Acceptable Theory of Acceptance: Partial Structures and the General Correspondence Principle. In French and Kamminga (1993). pp. 137–158.

da Costa, N.C.A. and S. French (1993b). A Model Theoretic Approach to "Natural Reasoning." *International Studies in Philosophy of Science* **7**, 177–190.

da Costa, N.C.A. and S. French (1995). Partial Structures and the Logic of the Azande. *American Philosophical Quarterly* **32**, 325–339.

da Costa, N.C.A. and S. French (forthcoming). *Partial Truth and Partial Structures: A Unitary Account of Models in Scientific and Natural Reasoning.* Book in preparation.

Dalla Chiara, M. L., K. Doets, D. Mundici and J. van Bentham, eds. (1997). *Structures and Norms in Science.* Dordrecht: Kluwer Academic Publishers.

Field, H. (1980). *Science without Numbers: A Defense of Nominalism.* Princeton, N.J.: Princeton University Press.

Field, H. (1982). Realism and Anti-Realism about Mathematics. *Philosophical Topics* **13**, 45–69. (Reprinted with a postscript in Field 1989, pp. 53–78.)

Field, H. (1988). Realism, Mathematics and Modality. *Philosophical Topics* **19**, 57–107. (Reprinted with changes in Field 1989, pp. 227–281.)

Field, H. (1989). *Realism, Mathematics and Modality.* Oxford: Basil Blackwell.

French, S. (1997). Partiality, Pursuit and Practice. In Dalla Chiara *et al.* (Eds.), 1997, pp. 35–52.

French, S. (forthcoming). Models and Mathematics in Physics: The Role of Group Theory. In Butterfield *et al.* (forthcoming).

French, S. and H. Kamminga, eds. (1993). *Correspondence, Invariance and Heuristics: Essays in Honour of Heinz Post.* Dordrecht: Reidel.

Hart, W.D., ed. (1996). *The Philosophy of Mathematics.* Oxford: Oxford University Press.

Hellman, G. (1989). *Mathematics without Numbers: Towards a Modal-Structural Interpretation.* Oxford: Clarendon Press.

Hellman, G. (1996). Structuralism without Structures. *Philosophia Mathematica* **4**, 100–123.

Hull, D., M. Forbes, and K. Okruhlik (Eds.), (1993). *PSA 1992: Proceedings of the 1992 Biennial Meeting of the Philosophy of Science Association*, Volume 2. East Lansing, Michigan: Philosophy of Science Association.

Ladyman, J. (1998). What is Structural Realism? *Studies in History and Philosophy of Science* **29**, 409–424.

Leonardi, P. and M. Santambrogio (Eds), (1993). *On Quine.* Cambridge: Cambridge University Press.

Mikenberg, I., N.C.A. da Costa and R. Chuaqui (1986). Pragmatic Truth and Approximation to Truth. *The Journal of Symbolic Logic* **51**, 201–221.

Putnam, H. (1967). Mathematics without Foundations. *Journal of Philosophy* **64**, 5–22. (Reprinted in Putnam 1979, pp. 43–59, and in Hart 1996, pp. 168–184.)

Putnam, H. (1979). *Mathematics, Matter and Method. Philosophical Papers*, Volume 1, Second edition. Cambridge: Cambridge University Press.

Resnik, M. (1981). Mathematics as a Science of Patterns: Ontology and Reference. *Noûs* **15**, 529–550.

Shapiro, S. (1989). Structure and Ontology. *Philosophical Topics* **17**, 145–171.

Shapiro, S. (1991). *Foundations Without Foundationalism: A Case for Second-Order Logic.* Oxford: Clarendon Press.

Shapiro, S. (1993). Modality and Ontology. *Mind,* **102**, 455–481.

Stapleton, T.J., ed. (1994). *The Question of Hermeneutics.* Dordrecht: Kluwer Academic Publishers.

Tait, W.W. (1986). Truth and Proof: The Platonism of Mathematics. *Synthese* **69**, 341–370. (Reprinted in Hart 1996, pp. 142–167.)

Tarski, A. (1983). The Concept of Truth in Formalized Languages. In *Logic, Semantic, Metamathematics.* Translated by J.H. Woodger. Second edition edited by John Corcoran. Indianapolis: Hackett. pp. 152–278.

van Fraassen, B.C. (1980). *The Scientific Image.* Oxford: Clarendon Press.

van Fraassen, B.C. (1985). Empiricism in the Philosophy of Science. In Churchland and Hooker (1985). pp. 245–308.

van Fraassen, B.C. (1989). *Laws and Symmetry.* Oxford: Clarendon Press.

van Fraassen, B.C. (1991). *Quantum Mechanics: An Empiricist View.* Oxford: Clarendon Press.

van Fraassen, B.C. (1993a). From Vicious Circle to Infinite Regress, and Back Again. In Hull *et al.* (1993). pp. 6–29.

van Fraassen, B.C. (1993b). Against Naturalized Empiricism. In Leonardi and Santambrogio (1993). pp. 66–88.

van Fraassen, B.C. (1994). Against Transcendental Empiricism. In Stapleton (1994). pp. 309–335.

Otávio Bueno
Division of History and Philosophy of Science
Department of Philosophy
University of Leeds
Leeds, LS2 9JT, UK
e-mail: phloab@leeds.ac.uk

Commentary by Chuang Liu

Any philosophical theory of mathematics, according to Paul Benacerraf, must meet the following *desiderata*: it must supply truth-conditions for all mathematical sentences and explain how we can have mathematical knowledge as we do. Benacerraf further argues that the only theory of truth which satisfies the former is a Tarski-type theory with a commitment to reference, and that of knowledge which fulfills the latter is one which allows the possibility of a causal access to the existing objects in question. Putting these together, either mathematical objects *actually* exist so that a Tarski-type truth theory can be had but such objects, being inert, are not causally accessible, or they do not actually exist (e.g., they are fictitious) so that no causal accessibility is required for knowing them, but then many mathematical terms will be without referents so that a Tarski-type truth theory is impossible. Hence, the Benacerraf dilemma for philosophy of mathematics, with which O. Bueno's article begins, and for which it aims to give a resolution.

For this to be a real dilemma, it seems, one would have to assume in addition that only a Tarski-type truth theory with actual objects in the domain of its relational structures may satisfy the first *desideratum* and the causal accessibility to such objects is required for knowing, i.e., having justified true beliefs about them. Moreover, the latter is necessary only if one is keen on an empiricist theory of mathematics, which is part of Bueno's approach. If so, an objection to either of these assumptions, if successful, would have been sufficed in avoiding the dilemma, but Bueno wants more; he wants to grab it by the horns and wrestle it to the ground: if one can have a theory of truth for M which does not involve actual objects for obtaining the satisfaction relations and which is robust enough to satisfy the first *desideratum*, one will be able to have knowledge in M without causal accessibility and thus fulfilling the second *desideratum*, where M is a domain of knowledge. And such a theory, Bueno argues, is supplied by that of partial structures and quasi-truth, a theory which has been developed recently by Newton da Costa and Steven French (among others).

The theory of quasi-truth avoids or grabs "the first horn of Benacerraf's dilemma" (depending on which you think is necessary for relieving the standoff) because on the one hand, it is a Tarski-type theory, so that it satisfies the first *desideratum*, and on the other, it does not necessitate a "commitment to mathematical objects." The latter is so precisely because it is a theory of *partial* truth or *quasi* truth. Now a sentence, Φ, is quasi-true in a partial structure S if and only if (i) S is a partial structure, (ii) there is a structure, B, which extends S to a full structure, and (iii) Φ is true in B, where being true (*simpliciter*) and a full structure are as defined in a standard Tarski theory (*cf.* Bueno 1997, p. 592). So, the theory is Tarski-type because after all Φ has to be true in B if it is quasi-true in S; and it does not necessarily make a commitment to Platonic entities in the domain of the structures in question because there does not have to be such entities in the domain of S (even though some of them may have to be assumed in B).[1] It is true that we may end up

---

[1] This may have put the point too crudely if not misleadingly. Since the structures A and B must share the same set of objects, those objects whose actual existence the empiricists such as Bueno want to deny must be found in other parts of B. This is, I guess, where structuralism becomes crucial to Bueno's theory, although the details of which are not yet clear. This is perhaps part of what Bueno said he would address on

with only quasi-truths in mathematics, but it is not necessarily a defect: why should we always ask for truth in mathematics if it is too much to do so in empirical sciences (empirical adequacy may be a better option if van Fraassen is right)? Hence something *"weaker than truth* should be countenanced as the aim of mathematics." So far so good.

Two questions (or types of questions) naturally arise with any proposal of the above kind: does it work, and is it the best alternative? Let us consider a very simple mathematical theory, T, which consists of a single sentence: 'the sum of two real numbers is a real number', or more precisely, 'for all x and y, if x and y are real numbers, then $z = f(x, y) = (x + y)$ is also a real number'. For its truth condition, the standard Tarski theory gives a model roughly as follows. It has a domain D of real numbers and a function $x \oplus y$ which maps any two members of D to a member and to which $f(\cdot, \cdot)$ refers. To avoid the second horn, one either devises a way in which none of the members of D must be understood as actual objects, or give up the Tarski-type truth theory completely (e.g., adopting a coherentist theory of truth). Clearly the first route is what Bueno takes, and let us now see if it works.

First, we are asked to see the analogy between the above and van Fraassen's *constructive empiricism* for theories of *empirical sciences*. Even though a theory, W, must be literally construed, i.e., its terms must genuinely refer, according to van Fraassen, it does not have to be true to be acceptable or adequate (*cf.* van Fraassen 1980). All we need and can ask for in science is empirical adequacy which concerns for each theory only those substructures that may enter into isomorphic relations with phenomena. Since truth is not the aim, some sentences in W which contain referring terms do not need to be true, and therefore no commitment is required to the entities to which those terms refer. They can be fictional objects whose sole use is to make the meaning of those sentences clear, just as objects in a computer simulation do. Then, we are asked to extend this picture to mathematics. But this can be difficult. To see the difficulty clearly, let us return to my pet mathematical theory T. What could possibly be the partial structure (which is an analog of van Fraassen's substructure) which on the one hand makes the given sentence partially true and on the other hand commits one to the existence of no causally inert objects (such as real numbers)? Note that van Fraassen has no difficulty of this kind with empirical theories. Let us say we have a theory, T', which consists of a single sentence: 'All tall men are bald', or more precisely, 'For all x, if x is a man and x is tall, x is bald'. The empirical adequacy of T' is given by having the sentence *true* in the standard Tarski sense of

---

another occasion.

truth, since *adequacy* may be *reduced* to *truth* when a commitment to all portended objects in a theory poses no problem for an empiricist. Such an option does not seem to be open to a partial-structure-quasi-truth account of mathematics. For numbers, either all of them are actual objects or none of them are; there does not seem to be a natural division among them in this respect that makes good (or common) sense. This is why most attempts in resolving the Benacerraf dilemma in the literature deal with numbers as a whole.

Of the previous attempts, Bueno mentions Hartry Field's nominalist approach, which clearly is an all-or-nothing affair with regard to numbers, and so are all other nominalist variations. Closer to home, there are the structuralists: Resnik, Shapiro and Hellman. The central idea is to replace numbers with relations or structures. Of course the question of the ontological status of numbers is transferred to that of structures given the technicality of the replacement is worked out. While Resnik and Shapiro defend a Platonist line, as Bueno pointed out, Hellman is a modalist, i.e., he considers mathematical structures to be possibilities only. These are also clearly all-or-nothing affairs. And all of these approaches if successfully worked out and defended may provide a resolution to Benacerraf's dilemma. Field's nominalism is obviously a way out. It is more difficult to see how the Platonist approaches of structuralism can accomplish the task. The truth theory is obviously of Tarski-type with all the referents being structures rather than some being numbers and some structures as in a non-structuralist variant. And one can also see how structures (relational properties) are less epistemically opaque as singular objects. And the escape provided by Hellman is that since accessibility to possible structures does not involve anything causal, the second horn of the dilemma is automatically avoided (or taken care of).

Because of the difficulties articulated above, I fail to see how Bueno's approach can be seen as clearly a better alternative to these existing ones. Towards the end of his article, Bueno seems to recommend a recipe which combines Hellman's modalism and the partial-structure-quasi-truth approach and which may be aptly called *partial modal structuralism*; the primary motivation for it being, if I am not mistaken, the desire to have an *empiricist* theory of mathematics, which parallels constructive empiricism in science. This is certainly an admirable attempt because if such a move succeeds, one can be an empiricist *simpliciter*. But to achieve this, it is not sufficient to carry the partial-structure-quasi-truth approach from science to mathematics. To make it plausible, one must first find the analogous *structure* (no pun intended) in mathematics as in empirical sciences which supports a clear separation of a partial structure form a full structure (or a X-normal structure). Only with such a structure in hand, can the above mentioned difficulty be adequately addressed.

REFERENCES

Bueno, O. (1997). Empirical Adequacy: A Partial Structures Approach. *Studies in the History and Philosophy of Science* **28**, 585–610.
Van Fraassen, B. (1980). *The Scientific Image.* Oxford: Clarendon Press.

Chuang Liu
Department of Philosophy
University of Florida
Gainesville, Florida 32611-2005
e-mail: cliu@phil.ufl.edu

Bueno's Reply to Liu

In the commentary above, C. Liu raises two main questions about the proposal I presented in "Empiricism, Mathematical Truth and Mathematical Knowledge:" (1) Does it work? (2) Is it the best alternative? According to Liu, the proper answer to both questions is *no*. In this reply, I shall spell out why I disagree, and argue for the opposite claim.

Let me start with (2). As is generally acknowledged, Benacerraf's (1973) problem provides a formidable difficulty for a philosophical account of mathematics (see Shapiro 1997, pp. 3–6; Field 1989, pp. 25–30; Hellman 1989, pp. 1–6; and Resnik 1997, pp. 82–87). After all, according to Benacerraf, the notions of mathematical truth and mathematical knowledge, if properly construed, cannot hang together. Since these two notions are also generally taken to be crucial for our understanding of mathematics, it is by no means surprising that so many proposals have tried to overcome the challenge.

As opposed to what is usually admitted, Liu claims that all extant views — in particular, those articulated by Field, Shapiro, Resnik, and Hellman — provide a resolution to Benacerraf's dilemma. If this were true, the dilemma would not be the provocative challenge it is, since it would be be met by whatever philosophy of mathematics we can think of — be it Platonist, nominalist or modalist. But I do think there is more to Benacerraf's difficulty than this. It may well be that the proposal I sketched in the paper above does not solve this problem after all, but it is certainly *not* the case that the existing proposals have all succeeded in this task.

In order to argue for this point, let me assemble the tentative solutions in two groups. On the one hand, the *timid* approaches are those which do not face Benacerraf's problem directly. They do not provide positive accounts of mathematical truth and mathematical knowledge, and in order to avoid Benacerraf's conclusion, they only *deny* some assumption or other in Benacerraf's argument. On the other hand, the *bold* approaches take the challenge at face value, and try to devise accounts of the two notions that, according to the challenge, cannot be articulated.

I do not think the timid approaches, even if successful in spotting a problematic hidden assumption in Benacerraf's argument, would settle the matter.[1] For the *costs* of their move is usually too high: if we relinquish the attempt to provide accounts of mathematical truth and mathematical knowledge, we can hardly make sense of mathematical practice, which is an activity embedded in the search for knowledge and truth.[2] In any case, without these notions, our philosophical understanding of mathematics is seriously jeopardized, since they arguably play a crucial role in the study of other problems in the philosophy of mathematics (such as the dynamics of mathematics and in mathematical heuristics).

Despite my great sympathies for Field's view, I think the account he provided is ultimately *timid.* In denying the existence of mathematical objects, and in resisting the indispensability argument with an ingenious nominalization program, Field was led to the claim that *there is no properly mathematical knowledge,* but only *logical* knowledge; that is, knowledge about what follows from what (see Field 1989, Chapter 3). But this conclusion is clearly at odds with mathematical practice, just as the related consequence that all existential mathematical statements are false (since there are no mathematical objects to satisfy them). Thus, without an account of mathematical knowledge, Benacerraf's dilemma is not solved. In my view, this provides the motivation for a proposal which

---

[1]Liu claims (in his paragraph 2) that Benacerraf's dilemma needs two of these assumptions: (a) "only a Tarski-type truth theory with actual objects in the domain of its relational structures may satisfy [Benacerraf's] first desideratum," that is, only this kind of theory can supply truth-conditions for all mathematical sentences, and (b) "causal accessibility to [mathematical] objects is required for knowing." However, rather than *assumptions* in Benacerraf's argument, these are *conclusions* for which Benacerraf has argued as part of his case (as Liu's himself acknowledges in paragraph 1). In fact, (a) and (b) are certainly *too strong* to be taken as assumptions in any persuasive dilemma — they require argument. Moreover, according to Liu, (b) is "necessary only if one is keen on an empiricist theory of mathematics." I would say exactly the contrary, since a crucial feature of van Fraassen's view is to *reject* causality as a decisive component of our understanding of science (see van Fraassen 1980, 1989). Thus, a *timid* empiricist approach to Benacerraf's problem could be articulated simply by *challenging* the role of causality in providing knowledge — that is, by criticizing (b).

[2]Notice that this claim does not make us immediately realists, since empiricists can provide appropriate formulations of the notions of mathematical truth and mathematical knowledge. This is what I have been trying to do.

is still nominalist — in the sense of *not requiring* the commitment to mathematical objects but which is *agnostic*, instead of atheist, about the existence of these objects. As is well known, van Fraassen (1980) countenances an agnosticism with regard to unobservable entities in science. I think a similar move should be taken by an empiricist philosophy of mathematics.

I also have a great sympathy for the proposal articulated by Hellman. In fact, as suggested at the end of my paper, an empiricist can adopt a modal-structural interpretation in order to run a nominalization program in mathematics — provided the interpretation is properly embedded into the partial structures framework. But, as opposed to Liu's claim, it is not at all clear that Hellman's account has settled Benacerraf's problem. At best, Hellman's strategy offers a *timid* answer, since as Liu correctly observes, possible structures are not taken to be subject to causal constraints. However, and more importantly, *no account of how we have knowledge of those structures is presented* (see also Shapiro 1997, pp. 228–229). And, as Hellman explicitly says, as to the "grounds for taking mathematical truth seriously . . . we shall be left hanging" (Hellman 1989, pp. 94–95). In other words, the two notions which are crucial to solve Benacerraf's difficulty are not addressed by the modal-structural interpretation.

My sympathies with the accounts developed by Shapiro and Resnik lie, of course, in their structuralism, since the proposal I advocate involves a crucial structuralist component. The divergencies are found at the ontological level, given the Platonist commitments incurred by these views. Shapiro and Resnik provide a *bold* move, adopting a standard Tarskian account of truth and a Platonist ontology. However, because of Platonism, it seems to me that their structuralist turn is not enough to guarantee a successful solution to Benacerraf's challenge.

Consider Shapiro's account. Our knowledge of mathematical structures is approached according to different strategies, depending on the complexity and size of the structures under consideration. Small, finite structures are grasped by a subject through abstraction via pattern recognition (Shapiro 1997, p. 11). However, given that those structures are abstract, the subject has no causal contact with them: "we grasp some structures through their systems, just as we grasp character types through their tokens" (p. 11). For more complex structures, Shapiro countenances extensions of pattern recognition. In particular, by noticing that some finite structures (the 2 pattern, the 3 pattern) come in sequence, the subject "extends the sequence to structures he has not seen exemplified" (p. 12). Eventually, he grasps the concept of a finite cardinal structure, and realizes that this structure instantiates in turn another structure: the

natural number structure — the first infinite pattern. Additional strategies are then presented to achieve still more complex structures: further
abstractions, the use of implicit definitions, and so on (pp. 112–142).

The problem here is that instead of an account of *how mathematical
knowledge is possible*, this view provides at most a *description* of mathematical techniques commonly used to refer to certain structures; rather
than an *explanation* of mathematical practice, we are offered a description of the data to be accommodated in the first place. But Benacerraf's
dilemma requires more: it demands a proper *explanation* of the data.
(Similar remarks apply to Resnik's account.)

Therefore, I conclude that Liu's claim that the extant approaches solve
Benacerraf's problem is unjustified.

But what about the proposal I sketched above: does it work? This is
Liu's question (1), and his negative answer is based on the following argument. In van Fraassen's view, there is a substantial distinction between
truth and empirical adequacy which ultimately derives from the distinction between observable and unobservable entities. Truth is equivalent
to empirical adequacy only if we restrict our consideration to observable
phenomena. According to van Fraassen, a scientific theory W is empirically adequate if there is an isomorphism between all the (empirical)
substructures of a given model of W and the (appropriately structured)
phenomena. Thus, in claiming that a scientific theory W does not have
to be true to be good, van Fraassen allows that sentences of W referring
to unobservables may not be true, thus avoiding ontological commitment
to unobservable entities.

But this move, Liu argues, is not open to a partial structures account
of mathematics, since "for numbers, either all of them are actual objects
or none of them are; there does not seem to be a natural division among
them" (p. 236). In other words, in mathematics, no distinction corresponds to the observable/unobservable demarcation in science. So, in
considering a mathematical theory, the account I suggested could only be
plausible once we find "the analogous *structure* . . . in mathematics as in
empirical sciences which supports a clear separation of a partial structure
from a full structure" (p. 236).

The problem with this argument is that it does not present the right
analogy between van Fraassen's account and the proposal I sketched. The
crucial aspect of a quasi-true theory is its *consistency* with an accepted
body of information about the world (represented by a body of accepted
sentences P). Since there are several different extensions of the information modeled in a partial structure A which are consistent with P,
no claim to the effect that a given theory is true is made. After all, in
some of these extensions, sentences of the theory can be false. Instead

of talking about observables and unobservables, we would rather consider the distinction between the *relations* that figure in a partial structure A and those which are involved in the corresponding A-normal extensions. This is the crucial structuralist move: what matters are *not* particular objects, but the *relations* we study. Now, Liu's argument enters at this point: what about the relations in the partial structure A, why are they acceptable to the empiricist in the first place? As Liu points out, if we are considering a mathematical theory, those relations already involve, say, numbers, and these are the sort of things that an empiricist cannot countenance.

It is precisely because of this that I think the empiricist needs a nominalization strategy. And this is why towards the end of my paper I suggested the adoption of a version of the modal-structural interpretation, appropriately recast in terms of partial structures. In this way, the empiricist can accommodate those partial structures needed to get off the ground: roughly speaking, it is enough to claim that these structures are *possible* (consistent with the body of information P). Moreover, the main difficulty faced by Hellman's formulation can then be overcome. For as opposed to Hellman's approach, the modal operators are not taken as primitive, but are spelled out in terms of quasi-truth. In this way, we have a unified account of mathematical and modal talk. Thus, the empiricist distinction between unproblematic ("observable") phenomena and problematic ("unobservable") features can be traced back to the distinction between the claim that a mathematical theory T is quasi-true at best (which does not commit us to the "truth" of the several A-normal extensions of a given partial structure) and the stronger, and problematic, assertion that T is indeed true. Just as in van Fraassen's account, the crucial move is to advocate an aim of mathematics which is weaker than truth.

Liu is certainly right in noticing that the motivation for the account I sketched is "the desire to have an *empiricist* theory of mathematics, which parallels constructive empiricism in science." Thus, he correctly maintains, "if such a move succeeds, one can be an empiricist *simpliciter*" (p. 236). I hope what I said has clarified the general outlook of this project, and the way I am trying to accomplish it.

REFERENCES

Benacerraf, P. (1973). Mathematical Truth. *Journal of Philosophy* **70**, 661–679.
Field, H. (1989). *Realism, Mathematics and Modality*. Oxford: Basil Blackwell.
Hellman, G. (1989) *Mathematics without Numbers*. Oxford: Clarendon Press.
Resnik, M. (1997). *Mathematics as a Science of Patterns*. Oxford: Oxford University Press.

Shapiro, S. (1997). *Philosophy of Mathematics: Structure and Ontology.* New York: Oxford University Press.
van Fraassen, B.C. (1980). *The Scientific Image.* Oxford: Clarendon Press.
van Fraassen, B.C. (1989). *Laws and Symmetry.* Oxford: Clarendon Press.

Poznań Studies in the Philosophy
of the Sciences and the Humanities
2000, vol. 71, pp. 243-260

Chuang Liu

COINS AND ELECTRONS: A UNIFIED
UNDERSTANDING OF PROBABILISTIC OBJECTS

ABSTRACT. By comparing two simple cases, the toss of fair coins (classical
objects) and the two-slit experiment of electrons (quantum objects), I suggest
a way of understanding the so-call wave-particle duality of quantum objects,
which deprives it of its paradoxical appearance. I argue that it is paradoxical
because we are misled into thinking that quantum objects *ought* to have categorical properties as classical corpuscles do, whereas many classical objects, if
characterized by their dispositional properties, can be just as puzzling. In the
end, I show how the distinction between categorical and dispositional properties may help to dispel the wave-particle paradox of quantum objects. And my
analysis, if correct, lends support for the position which takes disposition (and
propensity) seriously.

1.

Even though quantum theory (relativistic or not) is now indispensable for
physical sciences, when one asks what a quantum (or microscopic) object,
e.g. an electron, really is, it gives no clear answer. An electron is known
to be partly a corpuscle and partly a wave, but what does that mean? It
cannot mean that an electron consists of physically separate parts, one of
which acts like a particle and the other like a wave. What it is usually
taken to mean is that it *sometimes* acts like a particle and *sometimes* like
a wave. By acting like a particle or a wave, I mean no more than that the
object of interest possesses properties which are usually associated with
either particles or waves in classical physics. The properties in question
are, for particles, positions, momenta, energy, angular momenta (if they
are of finite extension), etc., and for waves (especially for wave packets),
extensions, amplitudes (intensities), wave lengths (or frequencies), group
velocities (if they are packets), etc. The properties do not have to be
observable; *they only have to be ones that an object can possibly possess.*

As to when quantum objects are waves and when they are particles,
the following brief account should suffice. When an electron is measured,
either by a Geiger counter or through a cloud chamber, the recorded effect

is such that whatever produces it must resemble a small particle, namely, that its total energy and momentum are delivered almost instantaneous to the measuring device over a very small region. A classical wave can never deliver its energy and momentum in such a manner; the total energy of a wave packet must be transferred to the device which receives it through the time that takes the packet to pass and over the region which matches the packet's size. Therefore, we *infer* that the electron must be a particle when measured. However, when not subject to measurement, it behaves like a wave (or wave packet). However, this statement needs careful explanation lest it is misunderstood (while misunderstandings of the wavelike nature of quantum objects abound).

As explained above, singular measurement outcomes do not indicate that electrons are like waves. However, if one looks at the collective outcome of a large number of measurements done on identical electrons and in identical environments, one sees it as consistent with each electron being wavelike.[1] In other words, "being wavelike," in this context, is a phrase that refers to *what* is capable of producing certain results in multi-particle experiments. The *inference* goes essentially as follows. The experiments are set up as such that if they were carried out on classical (or macroscopic) particles, they would result in identical outcomes. With identical electrons in such experiments, the outcomes are not identical but form a distribution of different individual outcomes whose total effect looks (when the details are ignored) exactly like what a classical wave would have left if measured in the same time interval. Since each single outcome is corpuscular, the wavelike behavior which is represented mathematically by the electrons' wave functions is no more than a property that produces probabilistic outcomes in an ensemble of identical electrons. Of course, identical electrons may not in fact be identical, but they are prepared in identical observable conditions and put through processes of identical observable properties and measured by identical apparatus and observers.

Electrons would not be considered puzzling if they can be viewed as corpuscles whose behavior is irreducibly probabilistic. The balls in a game of bagatelle board are objects of this kind,[2] and so are fair coins and dice. But the kind of probabilistic laws the electrons obey are fundamentally different from those obeyed by the classical objects, and they might be the result of some puzzling causal processes at the quantum level. I shall

---

[1]Here, the word 'identical' should be understood properly because, strictly speaking, no two electrons are identical. It means "being identical in all relevant aspects," where what is or is not relevant is determined by physical laws and the requirements of the experiments involved.

[2]We suppose that the game is played out under ideal conditions, so that the distribution at the bottom of the board is irreducibly probabilistic.

explain this point in the next section. The main challenge for this paper is to find a way of seeing this difference so that quantum objects, such as electrons, are not fundamentally different from classical ones, such as coins and dice. That they are usually seen as such is the result, I shall argue, of looking at the wrong set of properties when comparing them. In other words, it will be shown that if one insists on having an electron's states determined uniquely by the values of its observables, one gets the unsavory, or some even say paradoxical, conclusion: no electron (nor any other quantum object) has any definite property unless or until some measurement shows that it does.[3]

2.

Quantum objects are, nevertheless, not paradoxical in this respect, or so shall I argue in this paper: an electron is metaphysically no more mysterious or paradoxical than any ordinary object such as a fair coin or a die, provided that one accepts the idea that some classical objects are also irreducibly probabilistic. Electrons, for instance, appear puzzling only because we are misled by looking at them the wrong way: thinking that they *ought to be* like small charged particles simply because when they are measured (such as being put through a cloud chamber), they appear as such.

Let us now compare the tossing of fair coins with the two-slit experiment of electrons. Suppose that we have a fair coin which stays fair no matter how many times it is tossed and toss it repeatedly onto a hard and smooth surface (which stays hard and smooth no matter how many times it is hit by the coin) upon which it rests with either heads up or tails up (the only two possible outcomes of our experiment). Suppose again that the coin is tossed fairly or properly, meaning that the direction and speed by which it is tossed every time does not favor its landing on either heads or tails.[4] The probability of this coin's landing on heads under such tossing conditions is 0.5 and so is that of its landing on tails.

Let us assume that the probabilistic behavior of a fair coin when properly tossed is fundamental, namely, that no further specification of the physical conditions before and during the toss can change (i.e. increase or decrease) the probability of heads or tails. In other words, with a complete description of the states of the coin and its environment, the chance

---

[3]The same claim can be made with respect to local hidden-variables. For this, see the articles in Cushing and McMullin (1989).

[4]By properly tossing a fair coin, I mean the standard, if not too idealized, conditions under which the coin is tossed. Whether there is such a thing as the proper tossing of a fair coin and the probability of landing on heads or on tails in such tosses is 0.5 are issues I cannot discuss in this paper. Here they are taken for granted.

of its landing on heads or on tails are 50/50. One can of course change the probability by changing the set of tossing conditions: there must exist conditions under which all tosses of the coin result in heads up and there must be conditions under which all tosses result in tails up. What we have assumed here is that there is at least one set of conditions (i.e. fair or proper tossings of the coin) such that no description of which can improve on the estimation of the chance of the coin's landing heads or tails. By expressing the case in this way, I have certainly assumed that the coin's chance is objective under the given conditions. Beyond this objectivist prejudice on how the probabilities in this case should be interpreted, nothing more is assumed. Especially, I pass no ruling between the frequency and the propensity interpretation of probability. One can view the chance of the fair coin in question as the limiting frequency of an infinite *random* series defined as having maximal complexity (Martin-Löf 1966, 1969), or one can view it as propensity, i.e. some physical property possessed by the coin and the given set of conditions (Mellor 1971, Ch.4). To my best knowledge, both interpretations, which can also be subdivided into several different interpretations, have problems but are the best candidates for the objectivists (Hawson 1995).

However, one may reject the above assumption and insist on taking the coin-tossing experiment as deterministic in principle. In other words, if we toss the coin once and it falls on the surface heads up, then repeating the toss under exactly the same conditions will always make the coin falling heads up; and the same is true, *mutatis mutandis*, if the coin falls on the surface tails up. In this view, we have the distribution of 50/50 because we are not exactly repeating the same experiment, but doing different ones in which conditions producing either outcomes are randomly realized. In that case, we have no other choice but to abort the discussion. There is no point of arguing that a non-probabilistic object, if a fair coin is one, is not fundamentally different from a probabilistic one, such as an electron. Therefore, the reader may want to take the major claim of this paper to be a conditional: *if coins and electrons are both probabilistic objects*, they are not fundamentally different. And further I might add that there are good reasons for thinking that classical processes, such as the tossing of a fair coin or the throwing of a die, are irreducibly probabilistic (Earman 1984, Ch. 3).

We now turn to the case of an electron, a quantum object, and explain why it *appears* to be fundamentally different from a fair coin. In a two-slit experiment, a beam of electrons is produced by an oven such that one electron at a time is projected towards a measuring device: either a photo-sensitive surface or an array of Geiger counters. A thin wall is placed in between upon which two slits are cut in parallel, one directly above the other, so as to let the electrons through from two well separated openings. The outcomes of this experiment are dots or clicks that after a certain time interval form a distribution on the plate or among the Geiger counters. It is well known that the distribution is peculiar from a classical point of view. The probability of any of the electrons landing at a certain place on the photo-plate when the two slits are open simultaneously is not equal, as it should be if classical probability calculus is applicable, to the sum of the probabilities of its landing there when only the upper or only the lower slit is open, respectively. In short: $P_{oo} \neq P_{oc} + P_{co}$, where the symbol P_{oc} denotes the probability when the upper slit is open and the lower slit closed, P_{co} for the opposite situation, and P_{oo} when both slits are open.

To see why this is problematic or even mysterious, let us consider the following. Suppose that the electron in question is a particle, which means that it has certain position in space (and perhaps some other properties as well) at every instant, whether we know it or not. This implies that it passes through either the upper or the lower opening if it passes through the wall at all. In other words, if it has its position at certain instant right in the middle of the upper opening, then it cannot be at the middle of the lower one at that instant, and vice versa. But if this is true, then it follows that $P_{oo} = P_{oc} + P_{co}$, which contradicts the above result. This seems to imply that the opening of a slit which the given electron does not pass through has influence over where that electron may land. There is no apparent causal explanation for such an influence of an open slit upon an electron some distance away.[5] It is this kind of quantum phenomena that forces some pioneers of quantum theory to give up the idea that quantum objects possess, say, definite locations before they are observed to do so. Therefore, the alleged fundamental difference between a quantum object and a classically probabilistic one is that while it can be said that the latter possesses definite values at any instant of those properties which determine its state at that instant, albeit we may not or never know what they are, it cannot be said so for the former.

---

[5] One may think that the bit of material that covers the lower slit exerts some kind of causal influence upon the electron passing nearby so that the absence of which could alter the electron's later chance of landing somewhere on the plate. But quantum theory has no account of such a causal interaction. So, if it exists and is strong enough to alter the behavior of the electron, it must be some interaction that quantum theory fails to describe. Given quantum theory is complete, this is not possible, and if this is possible, quantum theory is radically incomplete. Bohmian mechanics is a theory that attempts to complete quantum theory. According to Bohmian mechanics, there is an interaction between the electron which passes through the upper slit and the physical condition at the lower slit, and it provides an account for such interactions. I can't discuss the status of Bohmian mechanics here, for that see Cushing *et al.* (1996).

Now, the conclusion reached above is only partially correct. To see how a different perspective should be given to this discussion which may dispel the mystery that quantum objects enjoys (they don't have definite properties until they are measured), let us take a closer look at the two cases again.

The two-slit experiment is only puzzling if we reason as follows. We see an electron impinging on the photo-plate with a precise position, and so we infer that it must or ought to have definite positions at every instant on its way to the plate. And when we find out that such inference breaks down because the mere assumption that it does have those positions (which may not be knowable to us) yields inconsistent statements, we conclude that it has no properties unless it is measured. It should not take a genius to see that the argument is fallacious. To be more clear and precise, observe the following. Isn't this the same as the experiment of the fair coin? We see the coin landing on the given surface with its heads facing up or down, and from that we should infer, by making the same kind of inference as in the electron case, that at every instant during its flight before it hits the surface, the coin has one of the two properties (heads up or heads down). But such an assumption would also lead us into inconsistent claims about the coin's behavior. Having such a property at each instant during its flight would contradict the predicted results of having the coin finally landing on the surface with a 50/50 chance of heads up or down. In this respect, the two cases — one of quantum, and the other of classical, objects — are indeed the same. If we are sure that in the case of tossing the fair coin, the coin is not having heads up or heads down at any instant of its movement towards the given surface, then we should be equally sure that the electron does not have any definite position at any instant of its movement towards the photo-plate. If nothing seems paradoxical or mysterious about the former, why should it be so for the latter?

One may argue that having a definite position (or energy or momentum) is intrinsic to an object, quantum or classical, while having heads up or down on a surface is not; therefore, the inference in the electron case is justified while the one in the case of a fair coin is not. We shall see later in Section 3 that this idea about which properties are intrinsic and which not is far from as simple as is assumed here. One may also argue that the coin is different from the electron in that we can calculate as well as "see" (with a quick enough video camera) coin's trajectory after it is tossed and before it hits the surface. That is certainly true, and we cannot say the same for the electron. However, having definite trajectories does not address the questions in the coin-tossing experiment, just as having a definite electric charge (and we know it) before the electron hits the measuring device does not help answer the questions in the two-slit experiment (nor does having a definite wave function of the electron at each instant).

Since the inference goes from the values of a certain property determined after a measurement to those of the same property before that, switching properties does not seem the right way to resolve the problem. Perhaps an argument can be made to relate different properties before and after a measurement, but this point cannot be clarified until we say more about different types of properties, which is the focus of Section 3. At present, the point is this. Since what is measured are the positions where the electrons hit the plate, having definite values for other properties is not relevant. The same is true with the coins. The outcomes of the toss are not the positions and the orientations of the coin, but whether its heads or tails face up when it settles on the surface. Nowhere in the process of the toss is a fair coin lying on a surface, so we must say that the coin does not possess definite values of that which is to be observed in the experiment. Furthermore, it would not help if one were to let a similar surface intersect the coin somewhere during the toss and find out whether or not it lands on its heads (and repeat this many times), because (a) if the experiment stops there, we still have the same problem; and (b) if one lets the experiment continue by releasing the coin again every time and letting it fall on the final surface, the probability distribution will not be the same (i.e. it is no longer a toss of a fair coin which has 50/50 chance of landing on its heads or tails).[6] Similarly, we may intersect an electron somewhere in the two-slit experiment with a Geiger counter (perhaps immediately behind one of the openings) and find out its definite position; and we either end the experiment there and face the same problem or let the electron go through and reach the photo-plate (or the Geiger counters), on which we will eventually (i.e. when we complete many runs) see a different probability distribution. In short, since what is relevant in both experiments is that which is to be observed in the end, neither kind of objects possesses precise or definite values for such properties in the middle of their respective experiment.

To summarize, if a fair coin's landing on heads or tails is a fundamentally probabilistic event and a coin is characterized by these two properties — heads up and tails up, a fair coin is just as mysterious (or nonmysterious) as an electron. The fact that we often choose to characterize a coin in some other way, for instance, by the trajectory of its center of mass and the orientations of its two faces in the air, either is beside the

---

[6]The distribution may incidentally end up the same as when the tosses are not intersected by a surface. But since it can only be incidental, the scenario does not affect the argument.

point or should prompt us to do the same with electrons. We automatically assume, because of our knowledge of classical objects, that electrons,
when not subject to measurement, should possess those observables whose
values are in fact only obtainable after the measurements. We now see
that if we make the same assumption to a fair coin, we would get an
equally puzzling object.

However, this is not to argue that we should characterize a tossed
fair coin only by the properties of its landing on heads or tails. On the
contrary, we should not do that. A tossed fair coin is better characterized
by its mass and trajectory in the air. But if, again, the probabilities are
fundamental, the trajectory plus the environment does not determine a
unique outcome of the toss. Is this not the same with an electron? It is
better characterized by its charge and wave function; but so characterized
we cannot determine which of its observable outcomes will obtain (we can
only determine each outcome's probability, just as in the case of the coin).
The only real difference between a fair coin and an electron in this respect
is that the trajectory of the former is itself an observable (we can "see" it)
or characterizable by observables, whereas what determines an electron's
states between measurements does not seem to be so characterizable. The
wave function has its range in the set of complex numbers and what it
represents (which physicists call a "wave packet"), if anything at all, is
not a wave in the four dimensional spacetime which one can set up an
experiment to measure. The "movement" represented by a wave function
of an electron, for instance, does not describe the electron's movement
in any straightforward sense; it mathematically determines, rather, (in a
complex space) the change of the probabilistic distribution of the possible
values of a given set of its observables. Therefore, what is observable over
time is the distribution not the wave.

3.

Even though physical objects are not identical to the collection of their
properties (as well as relations among them), the best way to settle questions regarding them is to examine those. An object may have one or both
of the two kinds of properties: *categorical* properties and *dispositional*
properties (for a discussion of categorical v. dispositional properties, see,
inter alia, Armstrong 1969, Mellor 1974, Pargetter and Prior 1982). One
may further divide the latter into *dispositions* and *propensities*, where the
latter are probabilistic versions of the former.[7] Categorical properties are
those occurrent or actual properties of an object whose values at a certain

---

[7]I shall use 'disposition' for deterministic dispositional properties and 'propensity'
for indeterministic dispositional properties. Similarly, a dispositional (propensional)

instant determine in part the object's state at that instant.[8] The values of such properties at one instant may be connected to their values at another instant by physical laws, but they are by no means determined by, or definable in terms of, those laws.[9] For many, a world, e.g. the actual one we inhibit, is determined by the totality of its categorical properties.[10] Dispositional properties are not occurrent; they are defined, rather, in terms of certain conditions or procedures that could produce certain occurrent properties. Suppose we have a property, P. *P is a dispositional property iff there are two sets of properties, A and B, such that if an object x were to possess A, it would possess B,* where A and B are sets of categorical properties.[11] Or similarly for propensity: *P is a propensity iff there are two sets of properties, A and B, such that if an object x were to possess A, it would possess B with a probability ρ. B* is usually known as the

---

object is an object which is primarily characterized by dispositional (propensional) properties.

[8] I assume here that all categorical properties are actual, but this might be controversial. On the one hand, there doesn't seem to be anything wrong with defining categorical properties as world-relative: each possible world has a set of properties "actually" obtaining in that world, but on the other hand, the very idea of any property obtaining in a possible world other than the actual one is defined counterfactually: property P obtains in world w rather than in the actual world means that if w were actual, property P (instead of some actually obtained property) would obtain. This seems to imply that an object can only have P in the actual world counterfactually; therefore, P itself is dispositional.

[9] In quantitative sciences, an object's properties and their values are usually distinguished as if the latter are not properties, but in fact that is not the case: they are all properties. For instance, we have a flag pole which is five feet long. Having length is one of the pole's properties and so is being five feet long. Therefore, the distinction is a pragmatic one for quantitative sciences which should not cause any philosophical confusion.

[10] For philosophers, these are called categorists (Armstrong 1983; Lewis 1985). Physicists usually do not state their commitment to categoricism explicitly, but it is taken for granted most of the time. For instance, anyone who regards the world as nothing more than the total collection of the worldlines in a 4-dimensional spacetime has to be a categorist. The values that uniquely determines a worldline are certainly values of categorical properties.

[11] Under most circumstances some laws are known to govern the behavior of an object, x, so that B is derivable from x's having A, but that may not always be the case. What is necessary is that the counterfactual statement, "if x were to have A then it has B," be true. Whether a counterfactual statement can be true without being supported by some laws is another issue of which I cannot pursue here. Moreover, there may be cases where B consists of properties that are not of x's. For instance, the properties that display the "penetrability" of a bullet (which is a common disposition) may belong to a sheet of glass or a block of wood, not at all to the bullet. Hence, a more inclusive definition of a dispositional property may be this: $\forall X$, *X is a dispositional property of x iff $\exists A, B$, such that A and B are categorical properties, and if x were to have A then B would necessarily be instantiated by some set of objects, where the "necessity" is physical necessity.*

display of *P*. Now, an categoricist is someone who believes in the priority
or the fundamentality of categorical properties.

The relations between these two kinds of properties are complex and
sometimes controversial. Categorical properties are usually taken as primary properties and dispositional ones as secondary, but such a view loses
its obviousness once it is scrutinized. It is possible, and even rigorously
argued by some, that all properties are dispositional. The idea is to regard
the obtaining of any property the result (and hence the display) of some
procedure by which its values can be measured, and to use that or the law
that governs it to define the property. For instance, a triangle's having
three angles, one of the most categorical of all categorical properties, can
be seen as a dispositional property (Mellor 1982) if having three angles is
the result of properly counting the angles of the triangle. Or the positions
of a cannon ball are determined and defined by the mechanical law that
governs its movement through space (Ellis and Lierse 1994). Of course,
the categoricists disagree; they regard the occurrent properties as prior or
fundamental, while laws are contingent relations (or connections) among
sets of those properties.

I have no intention to try to settle this debate, but the following fact
needs to be acknowledged: although all properties can be consistently
(and at least formally) rendered dispositional, the opposite is not true.[12]
No real dispositional property can be consistently construed as categorical.
It may be argued that this is not true because each disposition is definable
in terms of two sets of categorical properties (see the *A* and *B* in the above
definition) which are related by a counterfactual. But such a "definition"
has little consequence on the issues of categoricity: it does not imply
that the object possesses those categorical properties when it has the
disposition they define. Further, the very fact that a disposition is only
definable by a counterfactual shows that it is something over and above
the categorical properties in the definien (please ignore the use-mention
distinction here): it is some sort of lawlike connection. The *fragility* of
a vase or the *fairness* of a coin is a property that determines a lawlike
behavior (rather than the states) of the object such that knowing it one
knows not what the vase or the coin is but how it would behave were it
subject to certain conditions.

From another angle, it may also be argued that dispositions (and
propensities) are *grounded* in categorical properties because all dispositions either are *reducible* to or *supervene* on categorical properties of the
same object. For example, suppose that D is a set of mechanico-chemical

---

[12]I thank Kirk Ludwig who clarified for me matters discussed in this and the next
paragraph.

properties which make a vase fragile. Then, the vase breaks whenever it is thrown to a hard surface (*ceteris paribus*) because it has D. To generalize, it is plausible that for any disposition of an object, there is a set of categorical properties of the object that *makes* it, or *causes* it to, possess that disposition. In other words, x would exhibit B were it to have A because it has D, where A, B and D are sets of categorical properties which may not share any members. The relation between a disposition (or a propensity) and its categorical base (e.g. D) can be that of reduction or of supervenience, the discussion of whose differences and similarities has its own literature into which I have no time nor space to enter in this paper. The question that should be asked is whether the general thesis is true, namely, whether there is always a set of categorical properties to ground every disposition or propensity, whatever the nature of the grounding is. Some think that this is an empirical question that can only be answered via scientific investigations; and we shall see in a moment if it is so with our case.

Now, classical objects, such as fair coins and dice, have both categorical and propensional properties. It may be argued that the latter are grounded in the former, for example, a coin's propensity of landing heads or tails on a surface seems to be a groundable propensity. First of all, it seems safe to assume that the coin has definite categorical properties at every instant of its existence, i.e. where it is, how fast it moves or how its faces are oriented. Then, it seems equally obvious that the propensity in question supervenes on some of these properties in that no change of the probability of the coin's landing heads or tails can occur without a change of any of these properties. One may be able to generalize this result to all dispositional properties (probabilistic or not) of classical objects.

Let us now analyze the case of quantum objects. To cast the wave-particle duality into the categorico-dispositional scheme, we may say the following. The particle picture of an electron presents it as a *propensional object* (an object defined primarily by its propensional properties), i.e. what it would appear to be were it to be prepared and measured in a certain way. And the picture does not work in a categorical mode: not only can we not always infer from the known position of an electron after a measurement where it has been before that, but also it is inconsistent to assume, thanks to Bell's theorem, that it is definitely somewhere though we know not where.[13] The wave picture may be understood as presenting

---

[13] This is not true in one particular interpretation of quantum theory, namely, Bohm's interpretation (or Bohmian mechanics). But there is good reason to believe that Bohm's interpretation is not really an interpretation of quantum theory but rather a new theory for quantum objects (cf. note 5). It postulates new dynamical laws which are not present in the original quantum theory. Hence, my discussion in this paper

the electron as a *categorical object* (an object defined primarily by its categorical properties), except that the properties, such as the shape, movement and frequencies, etc., of the wave packet are of a peculiar type: they are not something one can directly measure with any apparatus.

Now, we can see why electrons and all quantum objects seem puzzling but are not: they do not have any classical (or observable) categorical properties that ground their propensities. But it does not mean that they do not have categorical properties at all which can do the grounding (in fact, if that is true, the states of a quantum object would be indeterminable, which is certainly not the case). If one denies that the wave function of an electron denotes a set of its categorical properties (e.g. its shape, movement and frequency) because of which it exhibits the propensities displayed in the measurement results, one must consider the electron as an *irreducibly propensional* object. Otherwise, one may opt to believe in the reality of electrons as wave packets and take the propensities to be grounded in their properties. Now, one must note that just because the dispositions of electrons are probabilistic (hence they are propensities), it does not mean that the grounding of them in the categorical properties is so as well. The propensities are uniquely determined (up to a set of phase transformations) by the wave functions. So, even the believers of the strong reduction thesis can have their case made if they can find ways to convince themselves that wave functions denote real categorical properties of quantum objects.[14]

Even if one has difficulty accepting the reification of wave functions, one does not necessarily have to give up categoricism (i.e. there are always categorical properties in which the propensities of quantum objects are grounded). The advance of physics may one day discover (by empirical means or not) some new categorical properties which may have a better ontological profile than wave functions and which may ground the propensities in a more satisfactory way.

To summarize, whether an object is irreducibly dispositional (or propensional) depends on whether there exits a set of its categorical properties which grounds all of its dispositions and/or propensities. It is if there doesn't. Some believe that this is an empirical question, but our analysis of quantum objects so far suggests otherwise. It is an empirical question

---

should be understood as being confined in quantum theory and its interpretations.

[14]To be more precise, the idea can be rendered as follows. Quantum objects are (ontologically, which also means categorically) identical to their wave packets which have categorical and genuinely wavelike properties, such as amplitudes, frequencies, etc. and which also uniquely (up to a phase factor) determine their physical states (this is also why they are categorical and intrinsic properties of the objects in question). Wave functions are mathematical representations of those properties; in other words, the extensions of wave functions are exactly these wavelike properties.

only if the grounding categorical properties must be observables, but I do not see why this must be so. Had observability been a necessary condition for introducing new categorical properties, none of the hidden variables theories would even have gotten off the ground.[15] Therefore, this is one of those places in science where metaphysical speculations seem inevitable. One may adopt an agnostic (or a positivist) view on this matter: I don't care unless a set of observable grounding categorical properties is found. This is itself a sort of metaphysical commitment.

I therefore conclude: both coins and electrons possess categorical as well as propensional properties. The latter may be grounded in the former, but the categorical properties that *ground* a propensity may not be the ones that *display* it. Those who infer from the positions of an electron after measurements to its having definite positions between measurements have confused these two sets of categorical properties: the positions after measurements display one of the electron's propensities but do not in this case ground it. While it is likely the case that all dispositions and propensities of classical objects are grounded in observable categorical properties, it is an open question whether the propensities of quantum objects are also so grounded. It is mostly likely that they are grounded but not in *observable* properties.

REFERENCES

Armstrong, D.M. (1983). *What Is a Law of Nature?* Cambridge: Cambridge University Press.

Armstrong, D.M. (1969). Dispositions are Causes. *Analysis* 30, 23–6.

Cushing, J.T., A. Fine and S. Goldstein (Eds.), (1996). *Bohmian Mechanics and Quantum Theory: An Appraisal.* Boston Studies in the Philosophy of Science. Dordrecht: Kluwer Academic Publishers.

Cushing, J.T. and E. McMullin, eds. (1989). *Philosophical Consequences of Quantum Theory.* Notre Dame: University of Notre Dame Press.

Earman, J. (1986). *A Primer on Determinism.* Dordrecht: D. Reidel.

Ellis, B. and C. Lierse (1994). Dispositional Essentialism. *Australasian Journal of Philosophy* 72, 27–45.

Hawson, C. (1995). Theories of Probability. *British Journal for the Philosophy of Science* 46, 1–32.

Lewis, D. (1985). A Subjectivist's Guide to Objective Chance. In D. Lewis, *Philosophical Papers,* Volume III. Oxford: Oxford University Press. pp. 83–113.

Martin-Löf, P. (1966). The Definition of Random Sequences. *Information and Control* 9, 602–619.

Martin-Löf, P. (1969). The Literature on von Mises' Kollectivs Revisited. *Theoria* 35, 12–37.

---

[15] For a brief but clear account of the hidden variables theories, see Redhead (1987, pp. 45–48). One of the essential constraints of a hidden variables theory is that from it no observational result be deducible that differs from the ones in quantum theory. If any hidden variable has independently observable values, the above constraint is violated. Therefore, no hidden variable can be an observable.

Mellor, D.H. (1971). *The Matter of Chance.* Cambridge: Cambridge University Press.

Mellor, D.H. (1982). Counting Corners Correctly. *Analysis* **42**, 96–97.

Mellor, D.H. (1974). In Defense of Dispositions. *The Philosophical Review* **83**, 157–81.

Pargetter, R.J. and E.W. Prior (1982). The Dispositional and the Categorical. *Pacific Philosophical Review* **63**, 366–70.

Chuang Liu
Department of Philosophy
University of Florida
Gainesville, FL 32611-8545
e-mail: cliu@phil.ufl.edu

Commentary by Steven French

Unification depends on similarity. In order to establish a unified understanding of coins and electrons, or classical and quantum objects in general, Professor Liu must show that they are relevantly similar. Similarity, in turn, may be viewed as a relationship between properties of objects. Thus Liu argues that coins and electrons have been regarded as fundamentally different only because we have been focusing on the wrong sets of properties. Shifting that focus, as Liu urges, will allow us to capture both sets of objects within the same metaphysical framework and provide a unified understanding of them.

In the first half of the paper it is argued that "an electron is metaphysically no more mysterious or paradoxical than any ordinary object such as a fair coin or die, provided that one accepts the idea that some classical objects are also irreducibly probabilistic" (p. 245). Well of course, one might respond, if the difference between classical and quantum physics hinges on the reducibility of probability, then abandoning it for classical objects is certainly going to blur the distinction between the two! However, Liu wants to penetrate a little deeper than this. He asks us to consider the two-slit experiment with electrons and then compare it with tossing a coin, nudging us towards the relevant similarities: just as inferring from observed to unobserved positions for the electron leads to inconsistency, so does the similar inference for coins when it comes to the property "heads up." If the coin is taken to possess such a property "in flight," then we lose the sense of objective chance we initially bought into. However, this seems to be little more than a restatement of our response above: a relevantly similar inconsistency arises in both cases only if we take the coins to be irreducibly probabilistic objects. It seems we haven't penetrated very far at all.

In the second half of the paper Professor Liu attempts to further dispel the mystery by arguing that quantum objects do not have observable categorical properties which ground their dispositional ones. From the perspective of the particle picture, an electron is seen to be a irreducibly "propensional object" in the sense that the propensities or dispositional properties determined by the wave function are not grounded in, or supervene on, any categorical properties (on pain of inconsistency). Alternatively, viewing the electron as wave-like, it can be regarded as "categorical" but the relevant properties, such as the frequency of the wave packet, cannot be regarded as directly measurable, unlike the case with coins.

Let us consider this latter option first. Regarding quantum objects as identical to their wave packets is an old idea with old pitfalls. In particular, once it is recalled that we are talking about functions in *configuration* space, it can be appreciated that when it comes to the categorical properties in this picture it is not just observability which is at issue. Indeed, if we consider a system of electrons and take the appropriate wave function as expressing the relevant categorical properties, then it seems that we must accept that such properties can no longer be regarded as possessed by the individual objects constituting, in some sense, the system but only by the latter itself taken as a whole. This is, of course, just the infamous quantum holism but it is precisely here, I suggest, that the fundamental differences with classical entities yawn wide.

Taking the alternative option of regarding the electron as irreducibly "propensional" might appear to be merely a case of sacrificing one form of metaphysical discomfort for another: instead of collections of objects possessing holistic categorical properties, we now have the objects themselves possessing ungroundable or non-supervenient dispositional properties. However, as Lie notes, such a view has already been defended within the classical context, so if there is a mystery here it is not one which cleaves the quantum from the classical.[1] And when it comes to a collection of such objects, the idea of non-supervenience is familiar through Teller's work (Teller 1989).

As is well known, Teller argued that we have inherited a metaphysical framework from classical physics which he calls "particularism" and which consists of the following three theses:

1. the world is composed of individuals

---

[1] Here I cannot resist a note of caution: to claim that "No *real* dispositional property can be consistently construed as categorical" (my emphasis) surely begs the question!

2. these individuals possess nonrelational properties

3. all relations between individuals supervene on such nonrelational properties (Teller 1989).

The "entangled" relations which hold between elements of a quantum
system cannot be fitted into this framework: if they are taken to supervene on "hidden" intrinsic properties of the particles, Bell's inequalities
result, and if they supervene on "extrinsic" properties the incorrect quantum state is obtained. Hence we must give up "particularism" in favor
of "relational holism," according to which such entangled relations are
irreducible or non-supervenient.

The central idea in this account is identical to that which underlies the
view of quantum objects as propensional: the superposition represented
by the wave function is characterized by ". . . an independently identifiable
property with distinctive experimental implications for the . . . system as
a whole" (Teller 1986, p. 80). Furthermore, if dispositional properties are
taken to be relational and categorical ones as non-relational, then particularism implies the supervenience of the dispositional on the categorical.
Rejecting that supervenience, as Liu does in the particle picture, means
rejecting particularism and thus the propensional view can be situated
within a broader metaphysical framework.

It is here that a certain degree of commonality between classical and
quantum physics might be sought, at least at the philosophical level.
Leaving aside the problematic case of spatio-temporal relations, physics
appears to have been wholly concerned with relations which are superve-
nient. Consider the way in which relations between objects tagged with
the names 'gravitational force' and 'electrostatic force' are shown to supervene on the masses and charges respectively. This has been an immensely
fruitful framework within which to operate but, as Teller argues, it may
no longer be appropriate.

According to Weyl our continued adherence to it may have its origin
". . . within the domain of sense data, which — it is true — can yield
but quality and not relation" (Weyl 1963, p. 4). Perhaps, then, it is an
unreflective form of empiricism which is at fault: we take the metaphysical
framework appropriate for the realm of sense data and attempt to carry
it with us when we ascend to the heights of physical theory. Teller's
suggestion is that we throw it away once we get to quantum mechanics.
Alternatively, we might abandon it even at the lower level: James, for
example, grounded his "radical empiricism" on the "statement of fact"
that ". . . the relations between things, conjunctive as well as disjunctive,
are just as much matters of direct particular experience, neither more nor
less so, than the things themselves (James 1932, p. xii)." It is in terms

of such metaphysical frameworks, I would urge, that we might achieve a unified understanding of quantum and classical objects.[2]

REFERENCES

French, S. (1989). Identity and Individuality in Classical and Quantum Physics. *Australasian Journal of Philosophy* **67**, 432–446.

James, W. (1932). *The Meaning of Truth*. London: Longmans, Green and Co.

Teller, P. (1986). Relational Holism and Quantum Mechanics. *British Journal for the Philosophy of Science* **37**, 71–81.

Teller, P. (1989). Relational Holism. In J. Cushing and E. McMullin (Eds.), *Philosophical Consequences of Quantum Theory*. Notre Dame: University of Notre Dame Press. pp. 208–235.

Weyl, H. (1949). *Philosophy of Mathematics and Natural Science*. Princeton: Princeton University Press.

Steven French
School of Philosophy
University of Leeds
Leeds, LS2 9JT, England
e-mail: s.r.d.french@leeds.ac.uk

Liu's Reply to French

I am very grateful that Professor French took the trouble to comment on my paper, which clarified points that ought to have been said more clearly and pointed out connections that ought to have been discussed or mentioned. I failed to discuss Teller's and also Healey's (1991) works on the particularism-holism issue partly because they are so well-known (I discussed them in another paper, see Liu 1996) and partly because I was too eager to find a way of *reorienting* our understanding of classical and quantum objects so as to relieve some of our "metaphysical angst" about them.

Professor French faulted me for assuming "irreducible probabilities" for classical objects, arguing that with that assumption, there won't be much of a difference to speak of between the classical and the quantum. I beg to differ on this point. The fundamental difference between the two hinges not on whether their probabilities are *reducible* but on whether their probabilities are *Boolean*. The "paradoxical" features of quantum objects

---

[2]In a sense I have tried to work towards this myself by considering the extent to which quantum objects can be regarded as (metaphysically) classical individuals (French 1989).

which generate our metaphysical anxiety are not from the irreducibility but from the *non-Boolean* character of quantum probabilities.

Professor French is entirely right in pointing out the difficulty of conceiving the wave packets in quantum mechanics as real categorical objects. At the moment I do not have a good argument in favor of such a move as much as I am convinced of it and have tried various arguments in the past. My hunch on this matter is that the fate of wave packets will be similar to that of, say, differentials which were once deemed incredible (by Berkeley?) as "departed quantities" (for "dx") and "the ghost of departed quantities" (for "d^2x"), and the "ghost of ghost . . ." (for higher order differentials).

REFERENCES

Healey, R. (1991). Holism and Nonseparability. *Journal of Philosophy* **88**, 391–421.
Liu, C. (1996). Holism vs. Particularism: A Lesson From Classical and Quantum Physics. *Journal for General Philosophy of Science* **27**, 267–279.

*Poznań Studies in the Philosophy
of the Sciences and the Humanities*
2000, *vol.* 71, *pp.* 261–277

Anna Maidens

ARE ELECTRONS VAGUE OBJECTS?

ABSTRACT. This paper examines and attempts to counter the claim that vagueness is always linguistic and there cannot be ontological vagueness. I look at a counter-example proposed by Lowe to Evans' "proof" that there cannot be vague objects. Using this counter-example, I argue that diachronic identity relations function in important ways like Lewis' counterpart relation in his semantics for modal logic. I raise the question of whether such relationships can be taken to be identity relations, but claim that recognizing their problematic nature does not necessarily lead to problems over the ontological status of their relata.

The claim that vagueness resides only in our language, and cannot be part of the world, is a common one in the literature (see, for example, Williamson 1994). In a famous one page article, Gareth Evans (1978) argued that there could not be vague objects, in the sense that identity statements involving singular terms referring to these objects might be indeterminate in truth value. More recently, Jonathan Lowe (1994) has argued against this conclusion. His argument hinges on a counterexample, involving imagining an electron which is captured by an atom to form an ion, which subsequently emits an electron. Lowe claims that it is ontologically indeterminate as to whether the electron which is captured is one and the same object as the electron which is emitted. I shall look at a possible worlds analogue of Lowe's case, and then argue that diachronic identity can be taken to function like Lewis' counterpart relation in his possible world semantics for modal logic. I shall then look both at Harold Noonan's (1995) reply to Lowe, in which he attempts to avoid the use of what he calls "identity invoking properties," and at Lowe's (1997) response, which I shall formalize in a tense logic in order to pursue the issue of vague diachronic identity.[1]

---

[1] In this paper I shall be interested in the extent to which Lowe's quasi-classical treatment of his example can be pursued. The characteristically quantum mechanical aspects of the situation (other than that of entanglement) will largely be ignored; a treatment of these aspects is given in Maidens (forthcoming), and the issues are treated at greater length in French, Krause and Maidens, "Quantum Vagueness" (unpublished).

1. Evans' Proof and Lowe's Diagnosis of the Flaw in the Proof

In summary, we consider singular terms 'a', 'b', assuming them to denote objects such that '$a = b$' is of indeterminate truth value. We introduce a sentential operator ∇ expressing the idea of indeterminacy.

(1) $\nabla(a = b)$.

Now introduce the property ascribed thus: "$\hat{x}[\nabla(x = a)]$" of its being indeterminate as to whether the thing possessing the property is identical with a.

(2) $\hat{x}[\nabla(x = a)]b$.

Further, assume that it is determinate as to whether '$a = a$' is true, so

(3) $\neg\nabla(a = a)$

(4) $\neg\hat{x}[\nabla(x = a)]a$

(5) $\neg(a = b)$ (Leibniz's law on (2) and (4))

There are a number of possible problems with this proof, but the ones I shall be concerned with involve the assumption that the singular terms 'a' and 'b' function rigidly (a point Lewis 1988 brings out) and a worry about whether the predicate introduced at line (2) legitimately describes a property.[2]

    Lowe diagnoses what he sees as the flaw in the proof in the following way, before offering the case of ionization as a specific counterexample to illustrate this flaw. In line (4) we ascribe to the object denoted by the name a the property denoted by $\hat{x}[\nabla(x = a)]$. However, Lowe claims, all we are able to ascribe to a is not the property of its being indeterminate whether it is identical with a (itself), but only the symmetric property

---

[2]Line (3) might be problematic if one denied that vague objects had *haecceities* or primitive thisniss; see for example Adams (1979), or if one thought the appropriate set-theory for the domains of semantic models for descriptions of vague objects was non-classical; see for example Krause (1992). Evans' version continues with the claim that the indeterminateness operator and its dual determinateness operator from a logic as strong as S5; but following Lewis' (1988) exegesis where indeterminateness and determinateness function like contingency and necessity, rather than possibility and necessity, Over (1989) has pointed out that they do not behave like the operators of S5. This can be seen quite easily by considering iterated operators: in S5 form "possibly P" we can derive both "possibly possibly P" and "necessarily possibly P"; the analogous expressions with 'determinately' and 'indeterminately' would contradict one another. Finally, Scott Shalkowski has suggested in conversation that the use of excluded middle may be problematic and the behavior of negation may be non-classical in vague contexts.

of its being indeterminate whether it is identical with b, $\hat{x}[\nabla(x = b)]$. Thus line (4) should read, $\hat{x}[\nabla(x = b)]a$, and we clearly cannot use this in conjunction with line (2) to generate a contradiction using Leibniz's law.

2. Counterparthood as an Analogy for Diachronic Identity

Lowe's suggested counterexample involves diachronic identity.[3] We are invited to imagine, let us say, a chlorine atom which in its neutral state contains 17 electrons. It captures an electron to form a chloride ion with 18 electrons, then re-emits an electron, returning to its neutral state. Why should we see the issue of identity as a problem? On what might be called the "classical" world view, there is no problem. We imagine (in a modification of Bohr's story) that the electrons within the atom circle round the nucleus in the manner of planets round the sun. Prior to ionization, the outermost orbit contains 5 electrons, in the manner of a rather sparsely populated asteroid belt.[4] On capture of an electron, for a brief period, there are 6, then the atom emits an electron, leaving 5 once more. On this classical story, with the electrons seen as akin to planets, clearly we simply have to follow the trajectory of the electron that is captured. Then either it will turn out that it is the one that is re-emitted, or it will turn out that it is not.

However, the quantum case is complicated by what is known as entanglement. The mathematical description of the composite of 6 electrons in the outer orbit or shell of the electron is such that we cannot (on the standard interpretation of quantum mechanics) follow the trajectory of any individual electron. Nor can we attach names or labels to the electrons when they are in this entangled state in any way which will enable us to track the electron which is captured to see if it is the one which is subsequently re-emitted. Yet, prior to capture, and after re-emission, there is no trouble naming or identifying the single free electron outside the atom. And since, during ionization, the number of electrons in the outer shell has increased from 5 to 6, it seems sensible to say that there is a sense in which the electron persisted through the process of ionization, and may have been re-emitted, or may have been left within the atom. Thus, it seems that in this rather curious case, it is genuinely indeterminate as to whether the electron which is re-emitted is identical with the electron which was previously captured.

---

[3]Sainsbury (1989) also raises the issue of vagueness in tensed and in modal identity statements.

[4]This story involves a monstrous hybrid of a quasi-classical picture no-one ever held, and modern physics which gives us the numbers in the outermost orbit (see, for example, White 1964, Appendix VI).

To analyze this curious case further, it is instructive to compare the issue of trying to establish relations of diachronic identity for electrons with an analogous case of trying to establish transworld identity relations (or counterpart relations, on a Lewisian account) across possible worlds. Consider first the set of time slices which constitute the history of the objects whose diachronic identity is at issue.

Slice 1: the electron a and atom prior to capture;

Slice 2: the ion after capture of the electron;

Slice 3: the electron b and atom after emission of the electron.

Compare this with the treatment Lewis offers of counterparthood, by envisaging the following set of possible worlds:

World 1: marble a outside box and box containing n marbles.

World 2: box containing $n + 1$ marbles.

World 3: marble b outside box and box containing n marbles.

This analogy is obviously missing some of the crucial features of the quantum case, most particularly that our n or $n + 1$ marbles are not in an entangled state (initially this disanalogy will not matter, and when it does become relevant, we shall return to the issue of entanglement in more detail). What are the possible answers to the question 'Is marble a the same marble as marble b'?

There are two ways of answering this question, which give opposite answers. On David Lewis' (1986) account, there is no trans-world identity, hence the answer is "no." Individuals appear in only one world, and modal statements about possibility, necessity and counterfactual situations are answered by reference to the counterpart in other possible worlds of the individual we are interested in within the actual world. Furthermore, it may well be the case that whether the marble outside the box in world 1 is the counterpart of the marble outside the box in world 3 is vague.[5] What is more, this vagueness seems to be ontic rather than epistemic: there is no single privileged relation of counterparthood between worlds (for reasons which will become apparent shortly). On Saul Kripke's account (1980),[6]

---

[5]One might ask why we don't just invoke Leibniz's Principle of the Identity of Indiscernibles at this point and collapse w_1 and w_3 to the same possible world; the answer lies, I think, in our beliefs about probabilities: see Kripke's story about the pair of dice.

[6]Strictly, this view is a combination of Kripke's approach to trans-world identity and a certain sort of modal realism: see Lewis (1986, pp. 198–9).

on the other hand, the answer to the question is "yes." There may be an epistemic problem (suppose all the marbles share all their intrinsic properties), but at an ontological level, once we have rigidly designated an individual by naming it 'a' we have picked out that individual (which in contrast to Lewis' picture exists in other possible worlds) and there is then a determinate fact of the matter as to whether 'a' rigidly designates the same individual as 'b'.

Now let us return to Evans' proof and see what happens when we read '$=$' as pertaining either to modal statements or to tensed statements (on the latter reading the relation will become one of diachronic identity).[7] First take the modal case. On Kripke's account, the individuals are designated rigidly by the act of naming them and the proof seems to go through unproblematically. Lowe's attempt to finesse the proof fails because properties expressed by predicates such as $\hat{X}[\nabla(x = a)]$ can at best be epistemically indeterminate: we may not *know* whether Hesperus is Phosophorus, but there is a fact of the matter, and if the identity statement is true, it is necessarily true and true in all possible worlds. Thus we can ask both if b possesses the property expressed by $\hat{X}[\nabla(x = a)]$ and whether a does; we do not, on a Kripkean account, have to ask about the symmetric property expressed by $\hat{X}[\nabla(x = b)]$.[8]

On a Lewisian account things look very different. The crucial steps in the proof become:

Assumption 1. 'a' and 'b' name individuals in w_1 and w_3 respectively. Names need not, indeed should not be construed as being rigid designators, because no individual occurs in two worlds.

(M1) $\nabla(a$ is the counterpart in w_1 of b in $w_3)$

$\Rightarrow$ (M2) $\hat{X}[\nabla (x$ is the counterpart in w_3 of a in $w_1)]b$

At this stage we realize that we cannot simply substitute 'counterpart' for '$=$' throughout, because assumption 2 and line (3) must be left as they were in the original proof.

What of line (4)? While the property $\hat{X}$ gave us information about the counterparts in w_3 of the individual b in w_1, the next step in the proof is now seen to be a claim about the property of being self identical within a single world, so we need to express this with a new predicate:

$\Rightarrow$ (M4) $\neg\hat{Y}[\nabla(x = a$ in $w_1)]a$

---

[7]Lewis (1986) pursues this analogy himself in his comparison of endurance and "perdurance" with trans-world identity and counterparthood.

[8]Interestingly, Lewis' (1988) account diagnoses the flaw as connected with rigidity.

The argument cannot proceed from this stage because we now have two predicates, one expressed by (M2) which is true of b, and the other expressed by (M4) (which follows from line (3)) which is true of a. For the argument to go forward to the conclusion, we would need an alternative to line (M4). This alternative cannot be generated from line (3). It could come from Lowe's parity of reasoning argument (according to which, for the original version of the argument, we can assume that if a is such that it is indeterminate whether it is identical with b then b is such that it is indeterminate whether it is identical with a). This gives us a property which we may express thus:

(M4′) $\hat{X}'[\nabla(x$ is the counterpart in w_1 of b in $w_3)]a$

but (similar to Lowe's point concerning his parity of reasoning claim) this does not deliver to us the conclusion of the argument. What would deliver such a conclusion would be the following property:

(M4″) $\hat{X}''[\nabla(x$ is the counterpart in w_3 of a in $w_1)]a$

However, since a is an individual in w_1 it cannot possess such a property since (M4″) asserts of it that it possesses the property of being whichever individual in w_3 is the counterpart of itself in w_1 and by hypothesis objects can only exist in one world.

But the example seems to have delivered more than we bargained for. What we have is not a counterexample to Evan's proof that there can be no vague identity statements, but an argument which shows that the counterpart relation in Lewisian possible world semantics can be vague precisely because it is *not* an identity relation.[9] But we started from a puzzle case in quantum mechanics where we wanted to argue that it was vague as to whether a prior to ionization was identical with b after re-emission of the electron. What is interesting is the formal similarity between counterparthood and diachronic identity (a similarity Lewis himself pursues), and the worry we have unearthed, that counterparthood is not an identity relation, has historical precedents in the case of diachronic identity. Hume, for example, suggests that diachronic identity is not properly an identity relation. Locke, similarly, encounters considerable difficulty in dealing with diachronic identity where it applies to large composite objects whose properties (and component parts) can change with time.

---

[9]For Lewis, all vagueness is linguistic: "the only intelligible account of vagueness locates it in our thought and language" (1986, p. 212). The fact that my program is opposed to this thought does not cause any problems with regard to my adoption of Lewis' approach to possible world semantics.

To see how these worries come to the fore in the temporal case, let us pursue the consequences of introducing a relation of "diachronic identity," intended to be taken as a claim about ontic relations between objects rather than the epistemic connections we make between our concepts. We then investigate what the steps in Evans' proof look like under such a relation. The steps come out in much the same way as in our earlier account of the Lewisian counterpart relation.

(**T1**) $\nabla(a$ existing at t_1 is diachronically identical to b existing at $t_3)$

$\Rightarrow$ (**T2**) $\hat{X}[\nabla(x$ at t_3 is diachronically identical to a at $t_1)]b$

Assumption 2. '$a = a$' is determinately true, where '$=$' stands for synchronic, or strict identity, is left unchanged, as is step (3). However, once more at this stage it becomes hard to see how to motivate the next step necessary for the proof, namely

(**T4**) $\neg\hat{X}[\nabla(x$ at t_3 is diachronically identical to a at $t_1)]a$

For part of what is at issue is the question of whether a is identical to one of the electrons at t_3 — it could be that it is identical with b, it could be that it is identical with one of the electrons in the atom, or it could be the case that the electron a has ceased to exist. The only statement we can assert (again using Lowe's parity of reasoning line) is the following:

(**T4'**) $\hat{X}'[\nabla(x$ at t_1 is diachronically identical to b at $t_3)]a$

and as we have seen, we cannot apply Leibniz's law to (T2) and (T4') to obtain the required contradiction.

However, as with the analogous case of counterparthood, we seem to have established more than we bargained for, because if we introduce this new relation of diachronic identity, the proof fails precisely because diachronic "identity" is not an identity relation in the sense of obeying the standard logical requirements for identity relations.

If we take Locke's view as paradigmatic of the classical picture of how object behave, we see that strict identity through time presupposes continuity of spatio-temporal history. Where histories split, join or cross (for example, the various collections of constituent bits of the Ship of Theseus), strict identity fails, and only the improper diachronic identity relation seems applicable. At this stage it is tempting to follow Hume and argue that the vagueness of such a relation lies in how we choose to use it, and there is no structure of relations in the world which it serves to name. On such a reading, to be an object continuing through time would be to be subject to behavior describable in terms of standard identity relations. Two objects at different times related by a relationship which did not fit into the pattern of an identity relation would not be identical, they would be determinately different.[10] It is often held that the analogue within quantum mechanics of bifurcating or crossing spacetime trajectories for classical objects is participation of quantum objects in entangled states. It would then seem natural to extend the above criticism to quantum objects: if electron *a* cannot be both identical in the strict sense with electron *b and* yet be indeterminately identical (for Evans' proof seems to go through on this assumption) but can at best be indeterminately "diachronically identical" with it, surely what we have established is that *a* seemingly ceases to exist on being incorporated into the ion, and *b* comes into being on being emitted. One way of trying to escape from this conclusion is to look for an approach to ontology which breaks away from classical presuppositions (French, Krause, Maidens, forthcoming). But if one wants to stick with a classical picture of the world, it is useful to see if an alternative approach can be used to escape Evans' proof, by questioning the legitimacy of the predicate introduced at line (2).

3. Noonan's Reply, and Tensed Properties

This is, in part, what prompts Noonan (1995) to suggest that we reformulate our discussion in terms of non-identity invoking properties. He asks us to imagine the case where we transplant Brown's brain (where Brown was a corpulent man) into Brownson's body (where Brownson is thin). It is supposedly indeterminate as to whether Brownson is the same person as Brown and thus similarly indeterminate as to whether Brown is fat or thin after the transplant. Yet it is not indeterminate as to whether Brownson is fat or thin after the transplant, thus, by invoking Leibniz's law we see that it can not really be indeterminate as to whether Brownson is the same person as Brown; he determinately is not.

We can approach this with a view to attacking Noonan's conclusion by temporally indexing the properties being discussed, as Lowe suggests in his (1997).[11] We temporally index the property $\hat{X}$ which will have to

---

[10]This conclusion would not follow on a non-classical conception of identity. For example, in Krause's (1992) quasi-set theory, there are some pairs of object which are neither identical nor non-identical. Instead of the axioms for identity we have indistinguishability axioms, and identity is a derived concept which only holds in certain circumstances.

[11]See Le Poidevin (1991) for a discussion of the relative merits of four approaches to the formalization of descriptions of objects undergoing change: propositional operators; temporally indexed properties; quantification over temporally indexed objects (temporal parts); and adverbial approaches. I shall introduce my treatment via tensed properties and then move onto quantification over temporal parts.

be non-identity invoking, as are, for instance, the properties Lowe mentions, of being captured by the atom and of being emitted by the atom.[12] At first it seems as though we need not make this move to non-identity invoking properties. We seem able to avoid Evans' conclusion by temporally indexing the identity invoking properties. For some time t during the lifetime of the ion it seems that b has the property *at times later than t* of its being indeterminate as to whether it is identical with a, while a fails to have the property *at times earlier than t* of its being indeterminate as to whether it is identical with a, so we cannot derive a contradiction from Leibniz's law.

A consideration of rigidity shows us why we do need to avoid identity invoking properties. It might still be argued that provided that a and b are rigid, however, either a names an individual which persists through the ionization process, or it does not. If it does not persist through the ionization process, then it is determinately non-identical with b. If a does persist through the ionization process, then since it fails to have the property of its being indeterminate whether it is identical with a at *all* times at which it exists, again, it is determinately non-identical with b. But this suggests that rigidity might be the problem, not the time dependence of properties. However, we do wish to avoid begging the question either by assuming we can talk of identity invoking properties or by assuming that the electrons can be discussed in some timeless way. To avoid the first, we use the non-identity invoking properties and to avoid the second, we use a tensed logic which, for any given time, will quantify over "the stock of then-existing individuals," (Rescher and Urquhart 1971, p. 234) where these individuals will be temporal parts of objects.[13] We do this by introducing as domains of quantification which are sets with temporal indices: we shall be interested in two in particular, which we shall call $\Gamma^p_{t_1}$ of all the individuals existing up to the time t_1 at which the electron is captured, and $\Gamma^p_{t_3}$ of all the individuals existing from the time t_3 at which an electron is emitted.[14]

We shall adopt the strategy of making reference to membership of one of the sets Γ in the list of properties we predicate of the variable bound by the quantifiers. This will make clear the connection with the earlier discussion of transworld identity versus counterparthood, and show how

---

[12]Such an approach is given in Rescher and Urquhart (1971) to deal with cases where a classical thing like a growing tree changes its properties with time, where we would not want to say on the basis of Liebniz's law that the tree on Monday cannot be the same tree as the tree on Tuesday because it has lost a leaf.

[13]See Lewis (1986) and Le Poidevin (1991) for a defense of temporal parts.

[14]It could be argued that domains of sharply defined individuals cannot provide the basis of an appropriate semantics for language which describes vague objects. However, to pursue this idea here would take us too far away from the purpose of this paper.

the issue of rigidity can be made explicit. In addition to the existential and universal quantifiers, we shall use a Russellian definite description operator **I** and introduce the predicates C (is captured by the atom),[15] and E (is emitted by the atom).

We then attempt to reconstruct Evans' proof as follows.

(T*1) $\nabla \left(\mathbf{I}y \left(y \in \Gamma_{t_1}^p \& Cy \right) = \mathbf{I}z \left(z \in \Gamma_{t_3}^f \& Ez \right) \right)$

(T*2) $\hat{X} \left[\nabla \left(x = \mathbf{I}y \left(y \in \Gamma_{t_1}^p \& Cy \right) \right) \right] \mathbf{I}z \left(z \in \Gamma_{t_3}^f \& Ez \right)$

Now this property predicated at line (T*2) of our latest version looks as though it is identity invoking, but need not be read that way: it claims that the item of which the property is predicated is such as to be whatever thing falls under a certain description.

Assumption: "$a = a$" is determinately true.

(T*3) $\neg \nabla \left(\mathbf{I}y \left(y \in \Gamma_{t_1}^p \& Cy \right) = \mathbf{I}y \left(y \in \Gamma_{t_3}^f \& Cz \right) \right)$

(T*4) $\neg \hat{X} \left[\nabla \left(x = \mathbf{I}y \left(y \in \Gamma_{t_1}^p \& Cy \right) \right) \right] \mathbf{I}y \left(y \in \Gamma_{t_1}^p \& Cy \right)$

We now need to introduce, following Rescher and Urquhart, an overtly temporalized version of Leibniz's law. If we introduce the temporal realization operator R_t, such that for an arbitrary proposition p, '$R_t(p)$' is to be read 'proposition p is realized at time t',[16] then for Leibniz's law, we have:

$$x = y \leftrightarrow (\forall \phi)(\forall t)[R_t(\phi x) \equiv R_t(\phi y)].$$

But at this stage the proof founders, for the proposition expressed by (T*4) explicitly mentions only times up to t_1, and it is by no means clear that it is realized at all times. In fact, it is far from clear whether an expression like that at line (T*2) is well formed, for it claims that a particular thing at time t_3 falls under a description which holds at t_1 but is not guaranteed to hold at other times, and thus it is doubtful whether the expression can meaningfully ascribe a property to anything (it certainly cannot meaningfully ascribe a property to a temporal part).

---

[15] If this reference to the future history of the electron seems worrying, it could presumably be replaced by some more cumbersome description which only refers to the past history of the electron couched in terms of the electron initially emitted from the source and introduced into the experimental apparatus.

[16] If one has worries about propositional operators of this sort as Johnston (1987) has (see Le Poidevin 1991 for a response to Johnston), the same point can be made without the introduction of such operators.

Thus attempting to formalize the argument in tensed logic gives us insight into the worry I raised at the beginning in stating Evans' proof, by showing the way in which the properties he introduces are illegitimate.

The sets $\Gamma^p_{t_1}$ and $\Gamma^f_{t_3}$ can be seen as collections formed by sets Γ_t of items at instants t. If there are enduring objects (in Lewis' sense where enduring contrasts with perduring), presumably these collections could have some of the same members (their members would be objects rather than temporal parts of objects). If (because of traditional worries about Leibniz's Law) we opt for Lewisian perdurance, the sets do not overlap. The situation then becomes very similar to that of the options for possible world semantics. Do the sets literally have the some of their members in common — i.e. is there a genuine relation of diachronic identity or transworld identity? Or do they have temporal parts or other-worldly counterparts related to one another in ways which do not necessarily obey the usual logic of identity? In the latter case, we can take persisting objects to be collections of temporal parts. Now Lewis suggests that in general we rarely encounter pathological cases where diachronic identity does not obey the logic of identity; for the most part causal and continuity considerations will pick out preferred sequences of temporal parts as constituting a single temporally extended object (in contrast to the modal case where pathology is rife). What the quantum mechanical example show is that for vague objects there is no preferred way of picking out sequences of such parts. Suppose we pick out an object at an instant, e.g. we point to it out *now*. If there are different collections of temporal parts at some later time which can be considered as earlier and later parts of the part we have picked now, then it is indeterminate as to whether it is identical or not with any given one of them. Such objects will not admit of rigidly designating names, if the function of a name involves "the notion that we can isolate and maintain contact with [over some period of time], 'in thought', the individual objects over which we generalize" (Greg McCulloch, 1989).

The major disanalogy between such an account of temporal slices (sets of temporal parts) and a Lewisian account of possible worlds is that possible worlds are causally and spatio-temporally isolated one from another. This disanalogy works to our advantage in discussing the notion of vague objects. There will be objects for which the temporal part relation can be uniquely specified: the unchanging particles of the classical world-view, Locke's corpuscles. But even in classical physics there will be objects whose temporal parts cannot be uniquely picked out. Two waves moving along a string, encountering one another at the midpoint and then separating would be a case in point. It might be argued that what this indicates is that waves are not objects; objects are precisely those things for which we have definite criteria for their identity across time. But presumably there is a similar ambiguity in the way we apply diachronic identity to the case of the Ship of Theseus and it would be an odd (and I would claim, undesirable) metaphysics which did not count ships as objects. Similarly, because of the way electrons enter into and then emerge from entangled states there is an vagueness in the relation relating their temporal parts.[17] Thus without spatiotemporal continuity, we have objects which can be considered to consist of sequences of temporal parts, where the relation between temporal parts is vague and hence identity statements about these objects are similarly vague.[18]

REFERENCES

Adams, R. (1979). Primitive Thisness and Primitive Identity. *Journal of Philosophy* **76**, 5–26.

Evans, G. (1978). Can There be Vague Objects? *Analysis* **38**, 208.

Johnston, M. (1987). Is There a Problem about Persistence? *Proceedings of the Aristotelian Society, Supplementary Volume* **62**, 197–35.

Krause, D. (1992). On a Quasi-Set Theory. *Notre Dame Journal of Formal Logic* **33**, 402–411.

Le Poidevin, R. (1991). *Change, Cause and Contradiction*. Basingstoke: Macmillan.

Lewis, D. (1986). *On the Plurality of Worlds*. Oxford: Basil Blackwell.

Lewis, D. (1988). Vague Identity: Evans Misunderstood. *Analysis* **48**, 128–30.

Lowe, E.J. (1994). Vague Identity and Quantum Indeterminacy. *Analysis* **54**, 110–114.

Lowe, E.J. (1997). Reply to Noonan on Vague Identity. *Analysis* **57**, 88–91.

Maidens, A. (Forthcoming), Vague Identity and Semantics for Quantum Logics. To appear in T. Childers, P. Kolár and V. Svoboda, eds., *The LOGICA Yearbook*. Prague: Filosofia.

McCulloch, G. (1989). *The Game of the Name*. Oxford: Oxford University Press.

Noonan, H. (1995). E.J. Lowe on Vague Identity and Quantum Indeterminacy. *Analysis* **55**, 14–19.

Over, D. (1989). Vague Objects and Identity. *Analysis* **49**, 97–99

Rescher, N. and A. Urquhart (1971). *Temporal Logic*. Vienna: Springer-Verlag.

Sainsbury, R.M. (1989). What Is a Vague Object? *Analysis* **49**, 99–103.

White, H.E. (1964). *Introduction to Atomic and Nuclear Physics*. Princeton: Van Nostrand.

Williamson, T. (1994). *Vagueness*. London: Routledge.

---

[17]Lowe (1997) also pins the problem onto the existence of entangled states. He makes the crucial point that it is uncontentious that $n + 1$ electrons both exist throughout the story (where n is the number of electrons in the atom prior to ionization) — even in the entangled state, but argues that there is no fact of the matter as to which is emitted.

[18]I should like to thank my colleagues at Leeds, particularly John Divers, Steven French, Robin Le Poidevin, Scott Shalkowski, Peter Simons and Roger White, and also Decio Krause, for comments on earlier versions of this paper, and Jonathan Lowe for providing pre-prints of articles.

Anna Maidens
Interdisciplinary Research Centre in Polymer Physics
University of Leeds
Leeds, LS2 9JT, England
e-mail: phyam@phys-irc.leeds.ac.uk

Commentary by David Over

Is vagueness only a feature of language or is it also present in the world? Are only words vague or do some words also refer to vague objects? These are the questions underlying Anna Maidens' stimulating paper about Evans (1978), as she points out. Before I comment on the deeper philosophical issues she raises, I want to give my views on the sentential operators Evans uses.

The Definite and the Indefinite

Evans introduces modal sentential operators, $\triangledown$ and $\triangle$, which he reads informally as "indefinitely" and "definitely." Given such brief readings, these triangles could have two interpretations when applied to a proposition p. The first could be called the "that-interpretation" of triangles, under which the readings are expanded to 'It is indefinitely true that p' and 'It is definitely true that p'. A philosopher who thinks that vagueness is only linguistic, that there is no vagueness in the world, might doubt that the that-interpretation makes sense. Assuming that it does, however, we should surely conclude that 'It is definitely true that p' logically implies p. There is solid evidence that Evans adopted the that-interpretation, since he explicitly supposed that 'indefinitely' and its dual 'definitely' are operators that generate a "logic as strong as S5" (p. 262). He also claimed that $\triangle p$ is a strengthening of p: that $\triangle p$ logically implies p. But from his axiom (3), which states not indefinitely $(a = a)$, we may infer definitely not $(a = a)$ by his supposition about duality, and then the absurd conclusion not $(a = a)$ by his claim about strengthening. In short, not all Evans' statements about sentential operators in his attempted proof can be true (Over 1989).

The second interpretation of the triangles is much to be preferred. This might be called the "whether-interpretation." Under it, 'indefinitely p' and 'definitely p' should be expanded as 'It is indefinite whether or not p' and 'It is definite whether or not p'. Note here that the latter, $\triangle p$ in this interpretation, no longer logically implies p. Anyone who endorses the purely linguistic account of vagueness should be happy with this whether-interpretation. From that point of view, we can explain the triangles using ways of making precise the terms in p, i.e., sharpenings or precisifications of p (Thomason 1982, Lewis 1988). We can state that $\triangle p$ is true if and only if p is true in all sharpenings, and $\triangledown p$ is true if and only if p is true in some sharpenings and false in others.

Evans' problem with his sentential operators is not really serious: it can be solved by adopting the whether-interpretation and employing $(p \,\&\, \triangle\, p)$ in the places where Evans wanted to use only p. Here we get, I think, as close as we coherently can to what Evans actually said. He supposed that the logic of his triangles is as strong as S5, but that can only be the case if $(p \,\&\, \triangle\, p)$ expresses the equivalent of the S5 necessity operator. Then Evans' vague logic would be similar to an axiomatization of S5 in which the necessity operator is expressed by 'p is true and it is non-contingent whether or not p is true' (Montgomery and Routley 1966).

It is misleading to compare this vague logic too closely with even that kind of axiomatization of modal logic, as there is no equivalent of the actual world in the semantics of vague logic. But in any case, the real problem for Evans lies in his presupposition that the singular terms in his attempted proof are *precise designators*. Maidens' paper brings out this fact very well.

Precise Designators

Maidens introduces domains of quantification, Γ_{t1}^{p} and Γ_{t3}^{f}, which she says at first are sets of temporal parts of things at times t_1 and t_3. She also has predicates, C and E, for the properties of being captured by the atom in her example and of being emitted by it, respectively. Her next step is to use the definite description operator, I, to give:

$$(T^*1)\quad \triangledown(Iy(y \in \Gamma_{t1}^{p}\,\&\,Cy) = Iz(z \in \Gamma_{t3}^{f}\,\&\,Ez))$$

However, I want to ask how it can be indefinite whether or not the temporal part captured at t_1 is identical to the temporal part emitted at t_3. How can any sharpening make identical the temporal parts referred to by these definite descriptions? They are indexed by distinct times if the terms t_1 *and* t_3 are not themselves vague.

Maidens later appeals to Lewis' distinction between persistence, endurance, and perdurance (Lewis 1986). Persistence is the theoretically neutral term: an object is said to persist if it exists at different times. Depending on our philosophical view, we will speak of an object as enduring if we believe it persists by being wholly present at different times, or alternatively of an object as perduring if we believe it persists by having temporal parts. In the light of her later use of this distinction, I would suggest that her domains consist of persisting objects, with the first definite description referring to the persistent object that is captured by the atom at t_1, and the second to the persistent object that is emitted by the atom at t_3. If this is so, then I would accept (T^*1), but agree that there is a fallacy at her (T^*2), where the second definite description is moved out of the intensional context. The point is that neither definite description is a precise designator — the equivalent in a vague context of a rigid designator.

What Maidens paper brings out very well for me is the difficulty of constructing precise designators for use in discussions of diachronic identity. Evans' attempted proof is only effective against believers in vague objects who also think that these vague objects can be precisely designated. But presumably the existence of vague objects would make precision in language even harder to attain than if vagueness were purely linguistic (Over 1989). Any attempt to introduce precise designators to help in a classical debate about diachronic identity will tend to frustrate vagueness, not only in the predicates and demonstratives used, but also in temporal terms and those used to try to keep track of spatiotemporal continuity. This problem is arguably unsolvable if there is objective vagueness as well in some physical processes, like causal ones, and in some physical objects, like electrons.

REFERENCES

Evans, G. (1978). Can There be Vague Objects? *Analysis* **38**, 208.

Lewis, D. (1986). *On the Plurality of Worlds*. Oxford: Blackwell.

Lewis, D. (1988). Vague Identity: Evans Misunderstood. *Analysis* **48**, 128–130.

Montgomery, H. A. and F. R. Routley. (1966). Contingency and Non-Contingency Bases for Modal Logics. *Logique et Analyse* **9**, 318–328.

Over, D.E. (1989). Vague Objects and Identity. *Analysis* **49**, 97–99.

Thomason, R.H. (1982). Identity and Vagueness. *Philosophical Studies* **42**, 329–332.

David Over
School of Social and International Studies
Forster Building
University of Sunderland
Sunderland, SR1 3SD, England
e-mail: david.over@sunderland.ac.uk

Maidens Reply to Over

David Over's response to my paper is both instructive and stimulating.
It is instructive in that it fills out the logical technicalities I hint at in
footnote 3, and draws attention to a certain carelessness of expression in
my formulation of the tensed version of Evans' proof. Over is entirely correct when he says that I should at first remain neutral in my terminology
and speak of $\Gamma^p_{t_1}$, etc., as the sets of persisting objects existing at a time,
leaving it neutral as to whether the members of these sets are temporal
parts of perduring objects or wholly present enduring objects.

I then agree that when we apply these two ways of understanding
persistence, the first (temporal parts) leads us to conclude that (T*1) is
not well formed, whereas the second enables us to see that a scope fallacy is
committed at (T*2). As Over points out, it is the second diagnosis which
is of interest; it suggests that the defender of vague objects can avoid
Evans' proof by denying that vague objects can be precisely designated.

I want to follow this up by offering a few comments on the theory
of singular reference one might wish to adopt. The singular term which
causes problems at line (T*2) is a definite description. Perhaps, a critic
might argue, the problem is due to our expecting a description to function
rigidly, a mistake Kripke drew to our attention some years ago. The
problem is outline in similar terms by Sainsbury (1989), who argues that
in Evans' proof, the terms a and b cannot have their reference fixed by
descriptive means, for then we will never know whether any resulting
conclusion is due to linguistic vagueness. I presume the only alternative
to a descriptive account of singular terms is a Kripkean one.

My response is to ask how, on Kripke's account of naming, we could
precisely designate a vague object, were there such things. Take Sainsbury's example, which is the claim that, as a matter of fact rather than
linguistic usage, the mountain Snowdon is such as to be compositionally
vague (it is vague what its parts are). Sainsbury's account involves us
having to fix the referent of the name 'Snowdon'. Now, on a Kripkean
story, this is done via a causal story leading back to some initial act of
baptism. We have already seen that for vague objects the causal chain
may be problematic because of difficulties with diachronic identity. But
I want to suggest the initial act of baptism is problematic. The act of
baptism, it seems to me, can take two forms. One proceeds by offering
a description which uniquely picks out the object in question. We have
already seen that this will not work. The second proceeds by an act of
ostention.

Suppose I opt for the latter. I stand on the summit and utter the
words 'this mountain is called Snowdon'. It is far from clear that such

an utterance would precisely designate the mountain on which I stand. It certainly designates it precisely enough for some practical purposes, for example, picking it out as distinct from Glyder Fach. But there is no sense in which it picks out the object precisely, including conveying my grasp on its precise boundaries, for the point is surely that if Snowdon is a vague object, it has no precise boundaries.

The challenge then becomes one of saying how our names for vague objects function in such a way that most of the time we can make descriptions of the sort exemplified by 'Snowdon is not Glyder Fach'.

REFERENCES

Kripke, S. (1972). *Naming and Necessity.* Oxford: Blackwell.
Sainsbury, R.M. (1989). What Is a Vague Object? *Analysis* **49**, 99–103.

POZNAŃ STUDIES IN THE PHILOSOPHY OF THE SCIENCES AND THE HUMANITIES

Contents of back issues

VOLUME 1 (1975)

Main topics:
The Method of Humanistic Interpretation; The Method of Idealization; The Reconstruction of Some Marxist Theories.
(sold out)

VOLUME 2 (1976)

Main topics:
Idealizational Concept of Science; Categorial Interpretation of Dialectics.
(sold out)

VOLUME 3 (1977)

Main topic:
Aspects of the Production of Scientific Knowledge.
(sold out)

VOLUME 4 (1978)

Main topic:
Aspects of the Growth of Science.
(sold out)

VOLUME 5 (1979)

Main topic:
Methodological Problems of Historical Research.
(sold out)

VOLUME 6 (1982)

SOCIAL CLASSES ACTION & HISTORICAL MATERIALISM

Main topics:
On Classes; On Action; The Adaptive Interpretation of Historical Materialism; Contributions to Historical Materialism.
(sold out)

VOLUME 7 (1982)

DIALECTICAL LOGICS FOR THE POLITICAL SCIENCE
(Edited by Hayward R. Alker, Jr.)

VOLUME 8 (1985)

CONSCIOUSNESS: METHODOLOGICAL AND PSYCHOLOGICAL APPROACHES
(Edited by Jerzy Brzeziński)

VOLUME 9 (1986)

THEORIES OF IDEOLOGY AND IDEOLOGY OF THEORIES
(Edited by Piotr Buczkowski and Andrzej Klawiter)

VOLUME 10 (1987)

WHAT IS CLOSER-TO-THE-TRUTH?
A PARADE OF APPROACHES TO TRUTHLIKENESS
(Edited by Theo A.F. Kuipers)

VOLUME 11 (1988)

NORMATIVE STRUCTURES OF THE SOCIAL WORLD
(Edited by Giuliano di Bernardo)

VOLUME 12 (1987)

POLISH CONTRIBUTIONS TO THE THEORY AND PHILOSOPHY OF LAW
(Edited by Zygmunt Ziembiński)

VOLUME 20 (1990)

Jürgen Ritsert
MODELS AND CONCEPTS OF IDEOLOGY

VOLUME 21 (1991)

PROBABILITY AND RATIONALITY
STUDIES ON L. JONATHAN COHEN'S PHILOSOPHY OF SCIENCE
(Edited by Ellery Eells and Tomasz Maruszewski)

VOLUME 22 (1991)

THE SOCIAL HORIZON OF KNOWLEDGE
(Edited by Piotr Buczkowski)

VOLUME 23 (1991)

ETHICAL DIMENSIONS OF LEGAL THEORY
(Edited by Wojciech Sadurski)

VOLUME 24 (1991)

ADVANCES IN SCIENTIFIC PHILOSOPHY
ESSAYS IN HONOUR OF PAUL WEINGARTNER ON THE OCCASION OF
THE 60TH ANNIVERSARY OF HIS BIRTHDAY
(Edited by Gerhard Schurz and Georg J.W. Dorn)

VOLUME 25 (1992)

IDEALIZATION III: APPROXIMATION AND TRUTH
(Edited by Jerzy Brzeziński and Leszek Nowak)

VOLUME 26 (1992)

IDEALIZATION IV: INTELLIGIBILITY IN SCIENCE
(Edited by Craig Dilworth)

VOLUME 27 (1992)

Ryszard Stachowski

THE MATHEMATICAL SOUL.
AN ANTIQUE PROTOTYPE OF THE MODERN MATEMATISATION OF
PSYCHOLOGY

VOLUME 28 (1993)

POLISH SCIENTIFIC PHILOSOPHY:
THE LVOV-WARSAW SCHOOL
(Edited by Francesco Coniglione, Roberto Poli and Jan Woleński)

VOLUME 29 (1993)

Zdzisław Augustynek and Jacek J. Jadacki

POSSIBLE ONTOLOGIES

VOLUME 30 (1993)

GOVERNMENT: SERVANT OR MASTER?
(Edited by Gerard Radnitzky and Hardy Bouillon)

VOLUME 31 (1993)

CREATIVITY AND CONSCIOUSNESS.
PHILOSOPHICAL AND PSYCHOLOGICAL DIMENSIONS
(Edited by Jerzy Brzeziński, Santo Di Nuovo, Tadeusz Marek and
Tomasz Maruszewski)

VOLUME 32 (1993)

FROM ONE-PARTY-SYSTEM TO DEMOCRACY
(Edited by Janina Frentzel-Zagórska)

J. Frentzel-Zagórska, *Introduction.* **Part I: Theoretical Approaches** – Z. Bauman, *A Post-modern Revolution*; L. Holmes, *On Communism, Post-communism, Modernity and Post-modernity*; L. Nowak, *The Totalitarian Approach and the History of Socialism*; J. Pakulski, *East European Revolutions and 'Legitimacy Crisis'.* **Part II: The Transitional Period** – A. Czarnota, M. Krygier, *From State to Legal Traditions? Prospects for the Rule of Law after Communism*; M. Szabó, *Social Protest in a Post-communist Democracy: Taxi Drivers' Demonstration in Hungary*; Z. Bauman, *Dismantling Patronage State*; E. Mokrzycki, *Between Reform and Revolution: Eastern Europe Two Years after the Fall of Communism*; J. Frentzel-Zagórska, *The Road to Democratic Political System in Post-communist Eastern Europe.* **Part III: The Case of Yugoslavia** – R.F. Miller. *Yugoslavia: The End of the Experiment.*

VOLUME 33 (1993)

SOCIAL SYSTEM, RATIONALITY, AND REVOLUTION
(Edited by Marcin Paprzycki and Leszek Nowak)

Introduction. **On the Nature of Social System** – U. Preuss, *Political Order and Democracy. Carl Schmitt and his Influence*; K. Paprzycka, *A Paradox in Hobbes' Philosophy of Law*; S. Esquith, *Democratic Political Dialogue*; E. Jeliński, *Democracy in Polish Reformist Socialist Thought*; K. Paprzycka, *The Master and Slave Configuration in Hegel's System*; M. Godelier, *Lévi-Strauss, Marx and After. A Reappraisal of Structuralist and Marxist Tools for Social Logics*; K. Niedźwiadek, *On the Structure of Social System*; W. Czajkowski, *Social Being and Its Reproduction.* **On Rationality and Captivity** – M. Ziółkowski, *Power and Knowledge.* L. Nowak, *Two Inter-Human Limits to the Rationality of Man*; M. Paprzycki, *The Non-Christian Model of Man: An Attempt at Psychological Explanation*; R. Egiert, *Toward the Sophisticated Rationalistic Model of Man.* **On Social Revolution** – L. Nowak, *Revolution is an Opaque Progress but a Progress Nonetheless*; K. Paprzycka, M. Paprzycki, *How do Enslaved People Make Revolutions?*; G. Tomczak, *Is it Worth Winning a Revolution?*; K. Brzechczyn, *Civil Loops and the Absorption of Elites*; R. McCleary, *What Makes Marxist Historical Materialism Objective?*; G. Kotlarski, *Classes and Masses in Social Philosophy of Rosa Luxemburg.* **On Real Socialism** – E. Gellner, *The Civil and the Sacred*; W. Marciszewski, *Economics and the Idea of Information. Why Socialism must have Collapsed?*; L. Nowak, K. Paprzycka, M. Paprzycki, *On Multilinearity of Socialism*; A. Siegel, *The Overrepression Cycle in the Soviet Union. An Operationalization of a Theoretical Model*; K. Brzechczyn, *The State of the Teutonic Order as a Socialist Society.* **Discussions** – R. McCleary, *Socioanalysis and Philosophy*; W. Heller, *Methodological Remarks on the Public and the Private in Hannah Arendt's Political Philosophy*; K. Brzechczyn, *On Unsuccessful Conquest and Successful Subordination.*

VOLUME 34 (1994)

Izabella Nowakowa

IDEALIZATION V: THE DYNAMICS OF IDEALIZATIONS

Introduction; Chapter I: *Idealization and Theories of Correspondence*; Chapter II: *Dialectical Correspondence of Scientific Laws*; Chapter III: *Dialectical Correspondence in Science: Some Examples*; Chapter IV: *Dialectical Correspondence of Scientific Theories*; Chapter V: *Generalizations of the Rule of Correspondence*; Chapter VI: *Extensions of the Rule of Correspondence*; Chapter VII: *Correspondence and the Empirical Environment of a Theory*; Chapter VIII: *Some Methodological Problems of Dialectical Correspondence.*

VOLUME 35 (1993)

EMPIRICAL LOGIC AND PUBLIC DEBATE.
ESSAYS IN HONOUR OF ELSE M. BARTH
(Edited by E.C.W. Krabbe, R.J. Dalitz, and P.A. Smit)

Part I: Interpersonal Reasoning: Conflicts and Fallacies – T. Govier, *Needing Each Other for Knowledge: Reflections on Trust and Testimony*; J. Woods, 'Secundum quid' as a Research Programme; G. Nuchelmans, *On the Fourfold Root of the 'argumentum ad hominem'*; F.H. van Eemeren, R. Grootendorst, *The History of the 'argumentum ad hominem' Since the Seventeenth Century*; L.S. van Epenhuysen, *Debate in a Bermuda Triangle of Medical Ethics*; E.C.W. Krabbe, *Reasonable Argument and Fallacies in the Kok-*

Stekelenburg Debate; **Part II: Linguistic and Conceptual Tools** – J.D. North, *Some Weak Links in the Great Chain of Being*; A. Næss, *'You assert this?' An Empirical Study of Weight Expressions*; R. Wiche, *Gerrit Mannoury on the Communicative Functions of Negation in Ordinary Language*; J. Hoepelman, T. van Hoof, *Default and Dogma*; Ch. Goossens, *On the Logic of Nonmoral Commitment*; W. Marciszewski, *Arguments Founded on Creative Definitions*; R. Jorna, *Cognitive Science and Connectivism: Friend and Enemy or Move and Counter-Move, an Application of Empirical Logic*; **Part III: Dialectical Climates and Tempests** – R.H. Johnson, *Dialectical Fields and Manifest Rationality*; P. du Preez, *Reason Which Cannot Be Reasoned With: What Is Public Debate and How Does it Change?*; M.A. Finocchiaro, *Logic, Democracy, and Mosca*; P.A. Smit, *The Logic of Virtue and Terror*; G. van Benthem van den Bergh, *On Obstacles to Public Debate*; J.P. van Bendegem, *Real-Life Mathematics versus Ideal Mathematics: The Ugly Truth*; **Part IV: The Disempowernment of Woman: Strategies and Counter-Moves** – V. Songe-Moller, *The Road of Being and the Exclusion of the Feminine. An Analysis of the Poem of Parmenides*; R.J. Dalitz, *The Subjection of Women in the Contractual Society. An Analysis of Thomas Hobbes' Theory of Agreement*; H. Schröder, *Anti-Semitism and anti-Feminism Again: The Dissemination of Otto Weininger's* Sex and Character *in the Seventies and Eighties*; J.R. Richards, *Traditional Spheres and Traditional Logic*.

VOLUME 36 (1994)

MARXISM AND COMMUNISM: POSTHUMOUS REFLECTIONS ON POLITICS, SOCIETY, AND LAW
(Edited by Martin Krygier)

M. Krygier, *Introduction*; A. Flis, *From Marx to Real Socialism: The History of a Utopia*; P. Marciniak, *The Collapse of Communism: Defeat or Opportunity for Marxism in Eastern Europe*; J. Clark, A. Wildavsky, *Chronicle of a Collapse Foretold: How Marx Predicted the Demise of Communism (Although He Called It "Capitalism")*; L. Nowak, *Political Theory and Socialism. On the Main Paradigms of Political Power and Their Methodological and Historical Legitimation*; E. Mokrzycki, *Marxism, Sociology, and "Real Socialism"*; R. Bäcker, *The Collapse of Communism and Theoretical Models*; A. Zybertowicz, *Three Deaths an Ideology: The Withering Away of Marxism and the Collapse of Communism. The Case of Poland*; M. Krygier, *Marxism, Communism, and the Rule of Law*; A. Czarnota, *Marxism, Ideology, and Law*; G. Skąpska, *The Legacy of Anti-Legalism*; A. Sajo, *Law and the Legal Scholarship in the Happiest Barrack and Among the Hungry Liberated: Personal Recollections*.

VOLUME 37 (1994)

THE SOCIAL PHILOSOPHY OF AGNES HELLER
(Edited by John Burnheim)

J. Burnheim, *Introduction*; M. Vajda, *A Lover of Philosophy – A Lover of Europe*; P. Despoix, *On the Possibility of a Philosophy of Values. A Dialogue within the Budapest School*; M. Jay, *Women in Dark Times: Agnes Heller and Hannah Arendt*; J.P. Arnason, *The Human Condition and the Modern Predicament*; R.J. Bernstein, *Agnes Heller: Philosophy, Rational Utopia and Praxis*; Z. Bauman, *Narrating Modernity*; P. Beilharz, *Theories of History – Agnes Heller and R.G. Collingwood*; P. Wolin, *Heller's Theory of Everyday Life*; P. Harrison, *Radical Philosophy and the Theory of Modernity*; A.J. Jacobson, *The Limits of Formal Justice*; P. Murphy, *Civility and Radicalism*; P. Murphy, *Pluralism and Politics*; V. Camps, *The Good Life: A Moral Gesture*; L. Boella, *Philosophy Beyond the Baseless and Tragic Character of Action*; G. Márkus, *The Politics of Morals*; A. Heller, *A Reply to My Critics*; *The Bibliography of Agnes Heller*.

VOLUME 38 (1994)

IDEALIZATION VI: IDEALIZATION IN ECONOMICS
(Edited by Bert Hamminga and Neil B. De Marchi)

Introduction – B. Hamminga, N. De Marchi, *Preface*; B. Hamminga, N. De Marchi, *Idealization and the Defence of Economics: Notes Toward a History*. **Part I: General Observations on Idealization in Economics** – K.D. Hoover, *Six Queries about Idealization in an Empirical Context*; B. Walliser, *Three Generalization Processes for Economic Models*; S. Cook, D. Hendry, *The Theory of Reduction in Econometrics*; M.C.W. Janssen, *Economic Models and Their Applications*; A.G. de la Sienra, *Idealization and Empirical Adequacy in Economic Theory*; I. Nowakowa, L. Nowak, *On Correspondence between Economic Theories*; U. Mäki, *Isolation, Idealization and Truth in Economics*. **Part II: Case Studies of Idealization in Economics** – N. Cartwright, *Mill and Menger: Ideal Elements and Stable Tendencies*; W. Balzer, *Exchange Versus Influence: A Case of Idealization*; K. Cools, B. Hamminga, T.A.F. Kuipers, *Truth Approximation by Concretization in Capital Structure Theory*; D.M. Hausman, *Paul Samuelson as Dr. Frankenstein: When an Idealization Runs Amuck*; H.A. Keuzenkamp, *What if an Idealization is Problematic? The Case of the Homogeneity Condition in Consumer Demand*; W. Diederich, *Nowak on Explanation and Idealization in Marx's Capital*; G. Jorland, *Idealization and Transformation*; J. Birner, *Idealizations and Theory Development in Economics. Some History and Logic of the Logic Discovery*. **Discussions** – L. Nowak, *The Idealizational Methodology and Economics. Replies to Diederich, Hoover, Janssen, Jorland and Mäki*.

VOLUME 39 (1994)

PROBABILITY IN THEORY-BUILDING.
EXPERIMENTAL AND NON-EXPERIMENTAL APPROACHES
TO SCIENTIFIC RESEARCH IN PSYCHOLOGY
(Edited by Jerzy Brzeziński)

Part I: Probability and the Idealizational Theory of Science – M. Gaul, *Statistical Dependencies, Statements and the Idealizational Theory of Science*. **Part II: Probability – Theoretical Concepts in Psychology – Measurement** – D. Wahlstein, *Probability and the Understanding of Individual Differences*; B. Krause, *Modeling Cognitive Learning Steps*; D. Heyer, R. Mausfeld, *A Theoretical and Experimental Inquiry into the Relation of Theoretical Concepts and Probabilistic Measurement Scales in Experimental Psychology*. **Part III: Methods of Data Analysis** – T.B. Iwiński, *Rough Set Methods in Psychology*; W. Koutstaal, R. Rosenthal, *Contrast Analysis in Behavioral Research*. **Part IV: Artifacts in Psychological Research and Diagnostic Assessment** – D.B. Strohmetz, R.L. Rosnow, *A Mediational Model of Research Artifacts*; J. Brzeziński, *Dimensions of Diagnostic Space*.

VOLUME 40 (1995)

THE HERITAGE OF KAZIMIERZ AJDUKIEWICZ
(Edited by Vito Sinisi and Jan Woleński)

Preface; J. Giedymin, *Ajdukiewicz's Life and Personality*; K. Ajdukiewicz, *My Philosophical Ideas*; L. Albertazzi, *Some Elements of Transcendentalism in Ajdukiewicz*; T. Batóg, *Ajdukiewicz and the Development of Formal Logic*; A. Church, *A Theory of the Meaning of Names*; M. Czarniawska, *The Way from Concept to Thought. Does it Exist in Ajdukiewicz's Semantical Theory?*; E. Dölling, *Real Objects and Existence*; J. Giedymin, *Radical Conventionalism, Its Background and Evolution: Poincaré, Le Roy, Ajdukiewicz*; J.J.

Jadacki, *Definition, Explication and Paraphrase in the Ajdukiewiczian Tradition*; G. Küng, *Ajdukiewicz's Contribution to the Realism/Idealism Debate*; J. Maciaszek, *Problems of Proper Names: Ajdukiewicz and Some Contemporary Results*; W. Marciszewski, *Real Definitions and Creativity*; K. Misiuna, *Categorial Grammar and Ontological Commitment*; L. Nowak, *Ajdukiewicz and the Status of the Logical Theory of Natural Language*; A. Olech, *Several Remarks on Ajdukiewicz's and Husserl's Approaches to Meaning*; A. Orenstein, *Existence Sentences*; J. Pasek, *Ajdukiewicz on Indicative Conditionals*; R. Poli, *The Problem of Position: Ajdukiewicz and Leibniz on Intensional Expressions*; M. Przełęcki, *The Law of Excluded Middle and the Problem of Idealism*; H. Skolimowski, *Ajdukiewicz, Rationality and Language*; K. Szaniawski, *Ajdukiewicz on Non-deductive Inference*; A. Varzi, *Variable-Binders as Functors*; V. Vasyukov, *Categorial Semantics for Ajdukiewicz-Lambek Calculus*; H. Wassing, *On the Expressiveness of Categorial Grammar*; J. Woleński, *On Ajdukiewicz's Refutation of Scepticism; Bibliography of Ajdukiewicz; Bibliography on Ajdukiewicz (writings in western languages)*.

VOLUME 41 (1994)

HISTORIOGRAPHY BETWEEN MODERNISM AND POSTMODERNISM. CONTRIBUTIONS TO THE METHODOLOGY OF THE HISTORICAL RESEARCH
(Edited by Jerzy Topolski)

Editor's Introduction; J. Topolski, *A Non-postmodernist Analysis of Historical Narratives*; F.R. Ankersmit, *Historism, Postmodernism and Historiography*; D. Carr, *Getting the Story Straight: Narrative and Historical Knowledge*; W. Wrzosek, *The Problem of Cultural Imputation in History. Culture versus History*; J. Tacq, *Causality and Virtual Finality*; G. Zalejko, *Soviet Historiography as a "Normal Science"*; H. Mamzer and J. Ostoja-Zagórski, *Deconstruction of Evolutionist Paradigm in Archaeology*; N. Lautier, *At the Crossroad of Epistemology and Psychology: Prospects of a Didactic of History*; Teresa Kostyrko, *Remarks on "Aesthetization" in Science on the Basis of History*.

VOLUME 42 (1995)

IDEALIZATION VII: IDEALIZATION, STRUCTURALISM, AND APPROXIMATION
(Edited by Martti Kuokkanen)

Idealization, Approximation and Counterfactuals in the Structuralist Framework – T.A.F. Kuipers, *The Refined Structure of Theories*; C.U. Moulines and R. Straub, *Approximation and Idealization from the Structuralist Point of View*; I.A. Kieseppä, *A Note on the Structuralist Account of Approximation*; C.U. Moulines and R. Straub, *A Reply to Kieseppä*; W. Balzer and G. Zoubek, *Structuralist Aspects of Idealization*; A. Ibarra and T. Mormann, *Counterfactual Deformation and Idealization in a Structuralist Framework*; I.A. Kieseppä, *Assessing the Structuralist Theory of Verisimilitude*. **Idealization, Approximation and Theory Formation** – L. Nowak, *Remarks on the Nature of Galileo's Methodological Revolution*; I. Niiniluoto, *Approximation in Applied Science*; E. Heise, P. Gerjets and R. Westermann, *Idealized Action Phases. A Concise Rubicon Theory*; K.G. Troitzsch, *Modelling, Simulation, and Structuralism*; V. Rantala and T. Vadén, *Idealization in Cognitive Science. A Study in Counterfactual Correspondence*; M. Sintonen and M. Kiikeri, *Idealization in Evolutionary Biology*; T. Tuomivaara, *On Idealizations in Ecology*; M. Kuokkanen and M. Häyry, *Early Utilitarianism and Its Idealizations from a Systematic Point of View*. **Idealization, Approximation and Measurement** – R. Westermann, *Measurement-Theoretical Idealizations and Empirical Research Practice*; U. Konerding,

Probability as an Idealization of Relative Frequency. A Case Study by Means of the BTL-Model; R. Suck and J. Wienöbst, *The Empirical Claim of Probality Statements, Idealized Bernoulli Experiments and their Approximate Version*; P.J. Lahti, *Idealizations in Quantum Theory of Measurement*.

VOLUME 43 (1995)

Witold Marciszewski and Roman Murawski
MECHANIZATION OF REASONING IN A HISTORICAL PERSPECTIVE

Chapter 1: *From the Mechanization of Reasoning to a Study of Human Intelligence*; Chapter 2: *The Formalization of Arguments in the Middle Ages*; Chapter 3: *Leibniz's Idea of Mechanical Reasoning at the Historical Background*; Chapter 4: *Between Leibniz and Boole: Towards the Algebraization of Logic*; Chapter 5: *The English Algebra of Logic in the 19th Century*; Chapter 6: *The 20th Century Way to Formalization and Mechanization*; Chapter 7: *Mechanized Deduction Systems*.

VOLUME 44 (1995)

THEORIES AND MODELS IN SCIENTIFIC PROCESSES
(Edited by William Herfel, Władysław Krajewski,
Ilkka Niiniluoto and Ryszard Wójcicki)

Introduction; **Part 1. Models in Scientific Processes** – J. Agassi, *Why there is no Theory of Models?*; M. Czarnocka, *Models and Symbolic Nature of Knowledge*; A. Grobler, *The Representational and the Non-Representational in Models of Scientific Theories*; S. Hartmann, *Models as a Tool for the Theory Construction: Some Strategies of Preliminary Physics*; W. Herfel, *Nonlinear Dynamical Models as Concrete Construction*; E. Kałuszyńska, *Styles of Thinking*; S. Psillos, *The Cognitive Interplay Between Theories and Models: the Case of 19th Century Optics*. **Part 2. Tools of Science** – N.D. Cartwright, T. Shomar, M. Suarez, *The Tool-Box of Science*; J. Echeverria, *The Four Contexts of Scienctific Activity*; K. Havas, *Continuity and Change: Kinds of Negation in Scientific Progress*; M. Kaiser, *The Independence of Scientific Phenomena*; W. Krajewski, *Scientific Meta-Philosophy*; I. Niiniluoto, *The Emergence of Scientific Specialties: Six Models*; L. Nowak, *Antirealism, (Supra-) Realism and Idealization*; R.M. Nugayev, *Classic, Modern and Postmodern Scientific Unification*; V. Rantala, *Translation and Scientific Change*; G. Schurz, *Theories and Their Applications – a Case of Nonmonotonic Reasoning*; W. Strawiński, *The Unity of Science Today*; V. Torosian, *Are the Ethics and Logic of Science Compatible?* **Part 3. Unsharp Approaches in Science** – E.W. Adams, *Problems and Prospects in a Theory of Inexact First-Order Theories*; W. Balzer, G. Zoubek, *On the Comparision of Approximative Empirical Claims*; G. Cattaneo, M. Luisa Dalla Chiara, R. Giuntini, *The Unsharp Approaches to Quantum Theory*; T.A.F. Kuipers, *Falsification Versus Effcient Truth Approximation*; B. Lauth, *Limiting Decidability and Probability*; J. Pykacz, *Many-Valued Logics in Foundations of Quantum Mechanics*; R.R. Zapatrin, *Logico-Algebraic Approach to Spacetime Quantization*.

VOLUME 45 (1995)

COGNITIVE PATTERNS IN SCIENCE AND COMMON SENSE.
GRONINGEN STUDIES IN PHILOSOPHY OF SCIENCE,
LOGIC, AND EPISTEMOLOGY
(Edited by Theo A.F. Kuipers and Anne Ruth Mackor)

VOLUME 46 (1996)

POLITICAL DIALOGUE: THEORIES AND PRACTICE
(Edited by Stephen L. Esquith)

VOLUME 47 (1996)

EPISTEMOLOGY AND HISTORY. HUMANITIES AS A PHILOSOPHICAL PROBLEM AND JERZY KMITA'S APPROACH TO IT
(Edited by Anna Zeidler-Janiszewska)

A. Zeidler-Janiszewska, *Preface*. **Humanistic Knowledge** – K.O. Apel, *The Hermeneutic Dimension of Social Science and its Normative Foundation*; M. Czerwiński, *Jerzy Kmita's Epistemology*; L. Witkowski, *The Frankfurt School and Structuralism in Jerzy Kmita's Analysis*; A. Szahaj, *Between Modernism and Postmodernism: Jerzy Kmita's Epistemology*; A. Grzegorczyk, *Non-Cartesian Coordinates in the Contemporary Humanities*; A. Pałubicka, *Pragmatist Holism as an Expression of Another Disenchantment of the World*; J. Sójka, *Who is Afraid of Scientism?*; P. Ozdowski, *The Broken Bonds with the World*; J. Such, *Types of Determination vs. the Development of Science in Historical Epistemology*; P. Zeidler, *Some Issues of Historical Epistemology in the Light of the Structuralist Philosophy of Science*; M. Buchowski, *Via Media: On the Consequences of Historical Epistemology for the Problem of Rationality*; B. Kotowa, *Humanistic Valuation and Some Social Functions of the Humanities*. **On Explanation and Humanistic Interpretation** – T.A.F. Kuipers, *Explanation by Intentional, Functional, and Causal Specification*; E. Świderski, *The Interpretational Paradigm in the Philosophy of the Human Sciences*; L. Nowak, *On the Limits of the Rationalistic Paradigm*; F. Coniglione, *Humanistic Interpretation between Hempel and Popper*; Z. Ziembiński, *Historical Interpretation vs. the Adaptive Interpretation of a Legal Text*; W. Mejbaum, *Explaining Social Phenomena*; M. Ziółkowski, *The Functional Theory of Culture and Sociology*; K. Zamiara, *Jerzy Kmita's Social-Regulational Theory of Culture and the Category of Subject*; J. Brzeziński, *Theory and Social Practice. One or Two Psychologies?*; Z. Kwieciński, *Decahedron of Education (Components and aspects). The Need for a Comprehensive Approach*; J. Paśniczek, *The Relational vs. Directional Concept of Intentionality*. **The Historical Dimension of Culture and its Studies** – J. Margolis, *The Declension of Progressivism*; J. Topolski, *Historians Look at Historical Truth*; T. Jerzak-Gierszewska, *Three Types of the Theories of Religion and Magic*; H. Paetzold, *Mythos und Moderne in der Philosophie der symbolishen Formen Ernst Cassirers*; M. Siemek, *Sozialphilosophische Aspekte der Übersetzbarkeit*. **Problems of Artistic Practice and Its Interpretation** – S. Morawski, *Theses on the 20th Century Crisis of Art and Culture*; A. Erjavec, *The Perception of Science in Modernist and Postmodernist Artistic Practice*; G. Dziamski, *The Avant-garde and Contemporary Artistic Consciousness*; H. Orłowski, *Generationszugehörigkeit und Selbsterfahrung von (deutschen) Schriftstellern*; T. Kostyrko, *The "Transhistoricity" of the Structure of Work of Art and the Process of Value Transmission in Culture*; G. Banaszak, *Musical Culture as a Configuration of Subcultures*; A. Zeidler-Janiszewska, *The Problem of the Applicability of Humanistic Interpretation in the Light of Contemporary Artistic Practice*; J. Kmita, *Towards Cultural Relativism "with a Small 'r'"*; The Bibliography of Jerzy Kmita.

VOLUME 48 (1996)

THE SOCIAL PHILOSOPHY OF ERNEST GELLNER
(Edited by John A. Hall and Ian Jarvie)

J.A. Hall, I. Jarvie, *Preface*; J.A. Hall, I. Jarvie, *The Life and Times of Ernest Gellner*. **Part 1: Intelectual Background** – J. Musil, *The Prague Roots of Ernest Gellner's Thinking*; Ch. Hahn, *Gellner and Malinowski: Words and Things in Central Europe*; T. Dragadze, *Ernest Gellner in Soviet East*. **Part 2: Nations and Nationalism** – B. O'Leary, *On the Nature of Nationalism: An Appraisal of Ernest Gellner's Writings on Nationalism*; K. Minogue, *Ernest Gellner and the Dangers of Theorising Nationalism*; A.D. Smith, *History and*

Modernity: Reflection on the Theory of Nationalism; M. Mann, *The Emergence of Modern European Nationalism*; N. Stagardt, *Gellner's Nationalism: The Spirit of Modernisation?* **Part 3: Patterns of Development** – P. Burke, *Reflections on the History of Encyclopaedias*; A. MacFarlane, *Ernest Gellner and the Escape to Modernity*; R. Dore, *Soverein Individuals*; S. Eisenstadt, *Japan: Non Axial Modernity*; M. Ferro, *L'Indépendance Telescopée: De la Décolonisation a L'Impérialisme Multinational*. **Part 4: Islam** – A. Hammoudi, *Segmentarity, Social stratification, Political Power and Sainthood: Reflections on Gellner's Theses*; H. Munson, Jr., *Rethinking Gellner's Segmentary Analysis of Morocco's Aif^cAtta*; J. Baechler, *Sur le charisme*; Ch. Lindholm, *Despotism and Democracy: State and Society in PreModern Middle East*; H. Munson, Jr. *Muslism and Jew in Morocco: Reflections on the Distinction between Belief and Behaviour*; T. Asad, *The Idea of an Anthropology of Islam*. **Part 5: Science and Disenchantment** – P. Anderson, *Science, Politics, Enchantment*; R. Schroeder, *From the Big Divide to the Rubber Cage: Gellner's Conception of Science and Technology*; J. Davis, *Irrationality in Social Life*. **Part 6: Relativism and Universals** – J. Skorupski, *The Post-Modern Hume: Ernest Gellner's 'Enlightenment Fundamentalism'*; J. Wettersten, *Ernest Gellner: A Wittgensteinian Rationalist*; I. Jarvie, *Gellner's Positivism*; R. Boudon, *Relativising Relativism: When Sociology Refutes the Sociology of Science*; R. Aya, *The Empiricist Exorcist*. **Part 7: Philosophy of History** – W. McNeill, *A Swang Song for British Liberalism?*; A. Park, *Gellner and the Long Trends of History*; E. Leone, *Marx, Gellner, Power*; R. Langlois, *Coercion, Cognition and Production: Gellner's Challenge to Historical Materialism and Post-Modernism*; E. Gellner, *Reply to my Critics*; I. Jarvie, *Complete Bibliography of Gellner's Work*.

VOLUME 49 (1996)

THE SIGNIFICANCE OF POPPER'S THOUGHT
(Edited by Stefan Amsterdamski)

Karl Popper's Three Worlds – J. Watkins, *World 1, World 2 and the Theory of Evolution*; A. Grobler, *World 3 and the Cunning of Reason*. **The Scientific Method as Ethics** – J. Agassi, *Towards Honest Public Relations of Science*; S. Amsterdamski, *Between Relativism and Absolutism: the Popperian Ideal of Knowledge*. **The Open Society and its Prospects** – E. Gellner, *Karl Popper – The Thinker and the Man*; J. Woleński, *Popper on Prophecies and Predictions*.

VOLUME 50 (1996)

THE IDEA OF THE UNIVERSITY
(Edited by Jerzy Brzeziński and Leszek Nowak)

Introduction – K. Twardowski, *The Majesty of the University*. **I** – Z. Ziembiński, *What Can be Saved of the Idea of the University?*; L. Kołakowski, *What Are Universities for?*; L. Gumański, *The Ideal University and Reality*; Z. Bauman, *The Present Crisis of the Universities*. **II** – K. Ajdukiewicz, *On Freedom of Science*; H. Samsonowicz, *Universities and Democracy*; J. Topolski, *The Commonwealth of Scholars and New Conceptions of Truth*; K. Szaniawski, *Plus ratio quam vis*. **III** – L. Koj, *Science, Teaching and Values*; K. Szaniawski, *The Ethics of Scientific Criticism*; J. Brzeziński, *Ethical Problems of Research Work of Psychologists*. **IV** – J. Goćkowski, *Tradition in Science*; J. Kmita, *Is a "Creative Man of Knowledge" Needed in University Teaching?*; L. Nowak, *The Personality of Researchers and the Necessity of Schools in Science*. **Recapitulation** – J. Brzeziński, *Reflections on the University*.

VOLUME 51 (1997)

KNOWLEDGE AND INQUIRY: ESSAYS ON JAAKKO HINTIKKA'S EPISTEMOLOGY AND PHILOSOPHY OF SCIENCE
(Edited by Matti Sintonen)

M. Sintonen, *From the Science of Logic to the Logic of Science*. **I: Historical Perspectives** – Z. Bechler, *Hintikka on Plentitude in Aristotle*; M.-L. Kakkuri-Knuuttila, *What Can the Sciences of Man Learn from Aristotle?*; M. Kusch, *Theories of Questions in German-Speaking Philosophy Around the Turn of the Century*; N.-E. Sahlin, *'He is no Good for My Work': On the Philosophical Relations between Ramsey and Wittgenstein*. **II: Formal Tools: Induction, Observation and Identifiability** – T.A.F. Kuipers, *The Carnap-Hintikka Programme in Inductive Logic*; I. Levi, *Caution and Nonmonotic Inference*; I. Niiniluoto, *Inductive Logic, Atomism, and Observational Error*; A. Mutanen, *Theory of Identifiability*. **III: Questions in Inquiry: The Interrogative Model** – S. Bromberger, *Natural Kinds and Questions*; S.A. Kleiner, *The Structure of Inquiry in Developmental Biology*; A. Wiśniewski, *Some Foundational Concepts of Erotetic Semantics*; J. Woleński, *Science and Games*. **IV: Growth of Knowledge: Explanation and Discovery** – M. Sintonen, *Explanation: The Fifth Decade*; E. Weber, *Scientific Explanation and the Interrogative Model of Inquiry*; G. Gebhard, *Scientific Discovery, Induction, and the Multi-Level Character of Scientific Inquiry*; M. Kiikeri, *On the Logical Structure of Learning Models*. **V: Jaakko Hintikka: Replies**.

VOLUME 52 (1997)

Helena Eilstein
LIFE CONTEMPLATIVE, LIFE PRACTICAL.
AN ESSAY ON FATALISM

Preface. **Chapter One: Oldcomb and Newcomb** – 1. *In the King Comb's Chamber of Game*; 2. *The Newcombian Predicaments*. **Chapter Two: Ananke** – 1. *Fatalism: What It Is Not?*; 2. *Fatalism: What Is It?*; 3. *Fatalism and* a priori *Arguments*; 4. *Fatalism and 'Internal' Experience*; 5. *Determinism, Indeterminism and Fatalism*; 6. *Transientism, Eternism and Fatalism*; 7. *Fatalism: What It Does Not Imply?*. **Chapter Three: Fated Freedom** – 1. *More on Libertarianism*; 2. *On the Deterministic Concept of Freedom*; 3. *Moral Self and Responsibility in the Light of Probabilism*; 4. *Fated Freedom*. **Chapter Four: The Virus of Fatalism** – 1. *Fatalism and Problems of Cognition*; 2. *The Virus of Fatalism: Why Mostly Harmless?*

VOLUME 53 (1997)

–
Dimitri Ginev
A PASSAGE TO THE HERMENEUTIC PHILOSOPHY OF SCIENCE

Preface; **Introduction**: *Topics in the Hermeneutic Philosophy of Science*; **Chapter 0**: *On the Limits of the Rational Reconstruction of Scientific Knowledge*; **Chapter 1**: *On the Hermeneutic Nature of Group Rationality*; **Chapter 2**: *Towards a Hermeneutic Theory of Progressive Change in Scientific Development*; **Chapter 3**: *Beyond Naturalism and Traditionalism*; **Chapter 4**: *A Critical Note on Normative Naturalism*; **Chapter 5**: *Micro and Macrohermeneutics of Science*; *Concluding Remarks*.

VOLUME 57 (1997)

EUPHONY AND LOGOS
ESSAYS IN HONOUR OF MARIA STEFFEN-BATÓG AND TADEUSZ BATÓG
(Edited by Roman Murawski, Jerzy Pogonowski)

Preface. **Scientific Works of Maria Steffen-Batóg and Tadeusz Batóg** – *List of Publications of Maria Steffen-Batóg; List of Publications of Tadeusz Batóg;* J. Pogonowski, *On the Scientific Works of Maria Steffen-Batóg;* Jerzy Pogonowski, *On the Scientific Works of Tadeusz Batóg;* W. Lapis, *How Should Sounds Be Phonemicized?;* P. Nowakowski, *On Applications of Algorithms for Phonetic Transcription in Linguistic Research;* J. Pogonowski, *Tadeusz Batóg's Phonological Systems.* **Mathematical Logic** – W. Buszkowski, *Incomplete Information Systems and Kleene 3-valued Logic;* M. Kandulski, *Categorial Grammars with Structural Rules;* M. Kołowska-Gawiejnowicz, *Labelled Deductive Systems for the Lambek Calculus;* R. Murawski, *Satisfaction Classes – a Survey;* K. Świrydowicz, *A New Approach to Dyadic Deontic Logic and the Normative Consequence Relation;* W. Zielonka, *More about the Axiomatics of the Lambek Calculus.* **Theoretical Linguistics** – J.J. Jadacki, *Troubles with Categorial Interpretation of Natural Language;* M. Karpiński, *Conversational Devices in Human-Computer Communication Using WIMP UI;* W. Maciejewski, *Qualitative Orientation and Gramatical Categories;* Z. Vetulani, *A System of Computer Understanding of Text;* A. Wójcik, *The Formal Development of van Sandt's Presupposition Theory;* W. Zabrocki, *Psychologism in Noam Chomsky's Theory (Tentative Critical Remarks);* R. Zuber, *Defining Presupposition without Negation.* **Philosophy of Language and Methodology of Sciences** – J. Kmita, *Philosophical Antifundamentalism;* A. Luchowska, *Peirce and Quine: Two Views on Meaning.* S. Wiertlewski, *Method According to Feyerabend;* J. Woleński, *Wittgenstein and Ordinary Language;* K. Zamiara, *Context of Discovery – Context of Justification and the Problem of Psychologism.*

VOLUME 58 (1997)

THE POSTMODERNIST CRITIQUE OF THE PROJECT OF ENLIGHTENMENT
(Edited by Sven-Eric Liedman)

S.-E. Liedman, *Introduction;* R. Wokler, *The Enlightenment Project and its Critics;* M. Benedikt, *Die Gegenwartsbedeutung von Kants aufklärender Akzeptanz und Zurückweisung des Modells der Naturwissenschaft für zwischenmenschliche Verhältnisse: Verfehlte Beziehungen der Geisterwelt Swedenborgs;* S.-E. Liedman, *The Crucial Role of Ethics in Different Types of Enlightenment (Condorcet and Kant);* S. Dahlstedt, *Forms of the Ineffable: From Kant to Lyotard;* P. Magnus Johansson, *On the Enlightenment in Psycho-Analysis;* E. Lundgren-Gothlin, *Ethics, Feminism and Postmodernism: Seyla Benhabib and Simone de Beauvoir;* E. Kiss, *Gibt es ein Projekt der Aufklärung und wenn ja, wie viele (Aufklärung vor dem Horizont der Postmoderne);* E. Kennedy, *Enlightenment Anticipations of Postmodernist Epistemology;* L. Nowak, *On Postmodernist Philosophy: An Attempt to Identify its Historical Sense;* M. Castillo, *The Dilemmas of Postmodern Individualism.*

VOLUME 59 (1997)

BEYOND ORIENTALISM. THE WORK OF WILHELM HALBFASS AND ITS IMPACT ON INDIAN AND CROSS-CULTURAL STUDIES
(Edited by Eli Franco and Karin Preisendanz)

E. Franco, K. Preisendanz, *Introduction and Editorial Essay on Wilhelm Halbfass;* Publications by Wilhelm Halbfass; W. Halbfass, *Research and Reflections: Responses to my*

VOLUME 60 (1998)

MARX'S THEORIES TODAY
(Edited by Ryszard Panasiuk and Leszek Nowak)

Production of "Rational Reality" and the "Systemic Coercion"; J. Bidet, *Metastructure and Socialism*; T. Andreani, *Vers une Issue Socialiste à la Crise du Capitalisme*; W. Becker, *The Bankruptsy of Marxism. About the Historical End of a World Philosophy*; D. Aleksandrowicz, *Myth, Eschatology and Social Reality in the Light of Marxist Philosophy.*

VOLUME 61 (1997)

REPRESENTATIONS OF SCIENTIFIC REALITY
CONTEMPORARY FORMAL PHILOSOPY OF SCIENCE IN SPAIN
(Edited by Andoni Ibarra and Thomas Mormann)

Introduction – A. Ibarra, T. Mormann, *The Long and Winding Road to Philosophy of Science in Spain.* **Part 1: Representation and Measurement** – A. Ibarra, T. Mormann, *Theories as Representations*; J. Garrido Garrido, *The Justification of Measurement*; O. Fernandez Prat, D. Quesada, *Spatial Representations and Their Physical Content*; J.A. Diez Calzada, *The Theory-Net of Interval Measurement Theory.* **Part 2: Truth, Rationality, and Method** – J.C. García-Bermejo Ochoa, *Realism and Truth Approximation in Economic Theory*; W.J.Gonzáles, *Rationality in Economics and Scientific Predictions*; J.P. Zamora Bonilla, *An Invitation to Methodonomics.* **Part 3: Logics, Semantics and Theoretical Structures** – J.L.Falguera, *A Basis for a Formal Semantics of Linguistic Formulations of Science*; A. Sobrino, E. Trillas, *Can Fuzzy Logic Help to Pose Some Problems in the Philosophy of Science?*; J. de Lorenzo, *Demonstrative Ways in Mathematical Doing*; M. Casanueva, *Genetics and Fertilization: A Good Marriage*; C.U. Moulines, *The Concept of Universe from a Metatheoretical Point of View.*

VOLUME 62 (1998)

IN THE WORLD OF SIGNS
(Edited by Jacek Juliusz Jadacki and Witold Strawiński)

Introduction. *How to Move in the World of Signs.* **Part I: Theoretical Semiotics** – A. Bogusławski, *Conditionals and Egocentric Mental Predicates*; W. Buszkowski, *On Families of Languages Generated by Categorial Grammar*; K.G. Havas, *Changing the World – Changing the Meaning. On the Meanings of the "Principle of Non-Contradiction"*; H. Hiz, *On Translation*; S. Marcus, *Imprecision, Between Variety and Uniformity: The Conjugate Pairs*; J. Peregrin, P. Sgall, *Meaning and "Propositional Attitudes"*; O.A. Wojtasiewicz, *Some Applications of Metric Space in Theoretical Linguistic.* **Part II: Methodology** – E. Agazzi, *Rationality and Certitude*; I. Bellert, *Human Reasoning and Artifical Intelligence. When Are Computers Dumb in Simulating Human Reasoning?*; T. Bigaj, *Analyticity and Existence in Mathematics*; G.B. Keene, *Taking up the Logical Slack in Natural Languages*; A. Kertész, *Interdisciplinarity and the Myth of Exactness*; J. Srzednicki, *Norm as the Basis of Form*; J.S. Stepanov, *"Cause" in the Light of Semiotics*; J.A Wojciechowski, *The Development of Knowledge as a Moral Problem.* **Part III: History of Semiotics** – E. Albrecht, *Philosophy of Language. Logic and Semiotics*; G. Deledalle, *A Philosopher's Reply to Questions Concerning Peirce's Theory of Signs*; J. Deledalle-Rhodes, *The Transposition of Linguistic Sign in Peirce's Contributions to "The Nation"*; R. E. Innis, *From Feeling to Mind: A Note on Langer's Notion of Symbolic Projection*; R. Kevelson, *Peirce's Semiotics as Complex Inquiry: Conflicting Methods*; J. Kopania, *The Cartesian Alternative of Philosophical Thinking*; Xiankun Li, *Why Gonsung Long (Kungsun Lung) Said "White Horse Is Not Horse"*; L. Melazzo, *A Report on an Ancient Discussion*; Ding-fu Ni, *Semantic Thoughts of J. Stuart Mill and Chinese Characters*; I. Portis-Winner, *Lotman's Semiosphere: Some Comments*; J. Réthoré, *Another Close Look at the Interpretant*; E. Stankiewicz, *The Semiotic Turn of Breal's "Semantique"*; **Part IV: Linguistic** – K. Heger, *Passive and Other Voices Seen from an*

VOLUME 63 (1998)

IDEALIZATION IX: IDEALIZATION IN CONTEMPORARY PHYSICS
(Edited by Niall Shanks)

VOLUME 64 (1998)

PRAGMATIC IDEALISM. CRITICAL ESSAYS ON NICHOLAS RESCHER'S SYSTEM OF PRAGMATIC IDEALISM
(Edited by Axel Wüstehube and Michael Quante)

VOLUME 65 (1999)

THE TOTALITARIAN PARADIGM AFTER THE END OF COMMUNISM.
TOWARDS A THEORETICAL REASSESSMENT.
(Edited by Achim Siegel)

A. Siegel, **Introduction**: *The Changing Fortunes of the Totalitarian Paradigm in Communist Studies.* **On Recent Controversies Over The Concept Of Totalitarianism** – K. von Beyme, *The Concept of Totalitarianism – A Reassessment after the End of Communist Rule*; K. Mueller, *East European Studies, Neo-Totalitarianism and Social Science Theory*; L. Nowak, *A Conception that is Supposed to Correspond to the Totalitarian Approach to Realsocialism*; E. Nolte, *The Three Versions of the Theory of Totalitarianism and the Significance of the Historical-Genetic Version*; E. Jesse, *The Two Major Instances of Totalitarianism: Observations on the Interconnection between Soviet Communism and National Socialism.* **Classic Concept Of Totalitarianism: Reassessment And Reinterpretation** – J.P. Arnason, *Totalitarianism and Modernity: Franz Borkenau's "Totalitarian Enemy" as a Source of Sociological Theorizing on Totalitarianism*; A. Sölner, *Sigmund Neumann's "Permanent Revolution": A Forgotten Classic of Comparative Research into Modern Dictatorship*; F. Pohlmann, *The "Seeds of Destruction" in Totalitarian Systems. An Interpretation of the Unity in Hannah Arendt's Political Philosophy*; W.J. Patzelt, *Reality Construction under Totalitarianism: An Ethnomethodological Elaboration of Martin Draht's Concept of Totalitarianism*; A. Siegel, *Carl Joachim Friedrich's Concept of Totalitarianism: A Reinterpretation*; M.R. Thompson, *Neither Totalitarian nor Authoritarian: Post-Totalitarianism in Eastern Europe.*

VOLUME 66 (1999)

Leon Gumański

TO BE OR NOT TO BE? IS THAT THE QUESTION?
AND OTHER STUDIES IN ONTOLOGY, EPISTEMOLOGY AND LOGIC

Preface; *The Elements of a Judgment and Existence*; *Traditional Logic and Existential Presuppositions*; *To Be Or Not To Be? Is That The Question?*; *Some Remarks On Definitions*; *Logische und semantische Antinomien*; *A New Approach to Realistic Epistemology*; *Ausgewählte Probleme der deontischen Logik*; *An Attempt at the Definition of the Biological Concept of Homology*; *Similarity.*

VOLUME 67 (1999)

Kazimierz Twardowski

ON ACTIONS, PRODUCTS AND OTHER TOPICS IN PHILOSOPHY
(Edited by Johannes Brandl and Jan Woleński)

Introduction; Translator's Note; Self-Portrait (1926/91); Biographical Notes. **I. On Mind, Psychology, and Language:** *Psychology vs. Physiology and Philosophy (1897); On the Classification of Mental Phenomena (1898); The Essence of Concepts (1903/24); On Idio- and Allogenetic Theories of Judgment (1907); Actions and Products (1912); The Humanities and Psychology (1912/76); On the Logic of Adjectives (1923/27).* **II. On Truth and Knowledge:** *On So-Called Relative Truths (1900); A priori, or Rational (Deductive) Sciences and a posteriori, or Empirical (Inductive) Sciences (1923); Theory of Knowledge. A Lecture Course (1925/75).* **III. On Philosophy:** *Franz Brentano and the History of Philosophy (1895); The Historical Conception of Philosophy (1912); On Clear and Unclear Philosophical Style (1920); Symbolomania and*

Pragmatophobia (1921); *Address at the 25th Anniversary Session of the Polish Philosophical Society (1929/31)*; *On the Dignity of the University (1933)*. **Bibliography**.

VOLUME 68 (2000)

Tadeusz Czeżowski

KNOWLEDGE, SCIENCE AND VALUES. A PROGRAM FOR SCIENTIFIC PHILOSOPHY
(Edited by Leon Gumański)

L. Gumański, *Introduction*. **Part 1: Logic, Methodology and Theory of Science** – *Some Ancient Problems in Modern Form*; *On the Humanities*; *On the Method of Analytical Description*; *On the Problem of Induction*; *On Discussion and Discussing*; *On Logical Culture*; *On Hypotheses*; *On the Classification of Sentences and Propositional Functions*; *Proof*; *On Traditional Distinctions between Definitions*; *Deictic Definitions*; *Induction and Reasoning by Analogy*; *The Classification of Reasonings and its Consequences in the Theory of Science*; *On the so-called Direct Justification and Self-evidence*; *On the Unity of Science*; *Scientific Description*. **Part 2: The World of Human Values and Norms** – *On Happiness*; *How to Understand "the Meaning of Life"?*; *How to Construct the Logic of Goods?*; *The Meaning and the Value of Life*; *Conflicts in Ethics*; *What are Values?*; *Ethics, Psychology and Logic*. **Part 3: Reality–Knowledge–World** – *Three Attitudes towards the World*; *On Two Views of the World*; *A Few Remarks on Rationalism and Empiricism*; *Identity and the Individual in Its Persistence*; *Sensory Cognition and Reality*; *Philosophy at the Crossroads*; *On Individuals and Existence*. J.J. Jadacki, *Trouble with Ontic Categories or Some Remarks on Tadeusz Czeżowski's Philosophical Views*; W. Mincer, *The Bibliography of Tadeusz Czeżowski*.

VOLUME 69 (2000)

Izabella Nowakowa, Leszek Nowak

THE RICHNESS OF IDEALIZATION

Preface; **Introduction** – *Science as a Caricature of Reality*. **Part I: THREE METHODO-LOGICAL REVOLUTIONS** – *1. The First Idealizational Revolution.Galileo's-Newton's Model of Free Fall; 2. The Second Idealizational Revolution. Darwin's Theory of Natural Selection; 3. The Third Idealizational Revolution. Marx's Theory of Reproduction.* **Part II: THE METHOD OF IDEALIZATION** – *4. The Idealizational Approach to Science: A New Survey; 5. On the Concept of Dialectical Correspondence; 6. On Inner Concretization. A Certain Generalization of the Notions of Concretization and Dialectical Correspondence; 7. Concretization in Qualitative Contexts; 8. Law and Theory: Some Expansions; 9. On Multiplicity of Idealization.* **Part III: EXPLANATIONS AND APPLICATIONS** – *10. The Ontology of the Idealizational Theory; 11. Creativity in Theory-building; 12. Discovery and Correspondence; 13. The Problem of Induction. Toward an Idealizational Paraphrase; 14. "Model(s) and "Experiment(s). An Analysis of Two Homogeneous Families of Notions; 15. On Theories. Half-Theories. One-fourth-Theories. etc.; 16. On Explanation and Its Fallacies; 17. Testability and Fuzziness; 18. Constructing the Notion; 19. On Economic Modeling; 20. Ajdukiewicz, Chomsky and the Status of the Theory of Natural Language; 21. Historical Narration; 22. The Rational Legislator.* **Part IV: TRUTH AND IDEALIZATION** – *23. A Notion of Truth for Idealization; 24. "Truth is a System": An Explication; 25. On the Concept of Adequacy of Laws; 26. Approximation and the Two Ideas of Truth; 27. On the Historicity of Knowledge.* **Part V: A GENERALIZATION OF IDEALIZATION** – *28. Abstracts Are Not Our Constructs. The Mental Constructs Are Abstracts; 29. Metaphors and Deformation; 30. Realism. Supra-Realism and Idealization.* **REFERENCES** – *I. Writings on Idealization; II. Other Writings.*

VOLUME 70 (2000)

QUINE. NATURALIZED EPISTEMOLOGY, PERCEPTUAL KNOWLEDGE AND ONTOLOGY
(Edited by Lieven Decock and Leon Horsten)

SCREEN-BASED ART

Annette W. Balkema and Henk Slager (eds.)

Amsterdam/Atlanta, GA 2000. 185 pp.
(Lier en Boog Series 15)
ISBN: 90-420-0801-6   **Paper Hfl. 37,-/US-$ 16.-**
ISBN: 90-420-0811-3   **Bound Hfl. 100,-/US-$ 43.-**

In the 21st century, the screen - the Internet screen, the television screen, the video screen and all sorts of combinations thereof - will be booming in our visual and infotechno culture. Screen-based art, already a prominent and topical part of visual culture in the 1990s, will expand even more. In this volume, digital art - the new media - as well as its connectedness to cinema will be the subject of investigation. The starting point is a two-day symposium organized by the Netherlands Media Art Institute Montevideo/TBA, in collaboration with the *L&B (Lier en Boog)* series and the Amsterdam School of Cultural Analysis (ASCA).

Issues which emerged during the course of investigation deal with questions such as: How could screen-based art be distinguished from other art forms? Could screen-based art theoretically be understood in one definite model or should one search for various possibilities and/or models? Could screen-based art be canonized? What are the physical and theoretical forms of representation for screen-based art? What are the idiosyncratic concepts geared towards screen-based art? This volume includes various arguments, positions, and statements by artists, curators, philosophers, and theorists. The participants are Marie-Luise Angerer, Annette W. Balkema, René Beekman, Raymond Bellour, Peter Bogers, Joost Bolten, Noël Carroll, Sean Cubitt, Călin Dan, Chris Dercon, Honoré d'O, Anne-Marie Duquet, Ken Feingold, Ursula Frohne, hARTware curators, Heiner Holtappels, Aernout Mik, Patricia Pisters, Nicolaus Schafhausen, Jeffrey Shaw, Peter Sloterdijk, Ed S. Tan, Barbara Visser and Siegfried Zielinski.

----------------------------------   *Editions Rodopi B.V.*

USA/Canada: 6075 Roswell Rd., Ste. 219, Atlanta, GA 30328, Tel. (404) 843-4314, *Call toll-free* (U.S.only) 1-800-225-3998, Fax (404) 843-4315

All Other Countries: Tijnmuiden 7, 1046 AK Amsterdam, The Netherlands. Tel. ++ 31 (0)20 6114821, Fax ++ 31 (0)20 4472979

THE MORAL STATUS OF PERSONS
Perspectives on Bioethics

Ed. by Gerhold K. Becker

Amsterdam/Atlanta, GA 2000. VII,246 pp.
(Value Inquiry Book Series 96)
ISBN: 90-420-1201-3   Hfl. 100,-/US-$ 55.50

The advances in molecular biology and genetics, medicine and neurosciences, in ethology and environmental studies have put the concept of the person firmly on the philosophical agenda. Whereas earlier times seemed to have a clear understanding about the moral implications of personhood and its boundaries, today there is little consensus on such matters. Whether a patient in the last stages of Alzheimer's disease is still a person, or whether a human embryo is already a person are highly contentious issues.
This book tackles the issue of personhood and its moral implications head-on. The thirteen essays are representative of the major strands in the current bioethical debate and offer new insights into humanity's moral standing, its foundations, and its implications for social interaction. While most of the essays approach the issue by drawing on the rich intellectual tradition of the West, others offer a cross-cultural perspective and make available for ethical consideration the philosophical resources and the wisdom of the East. The contributors to this book are highly recognized philosophers, ethicists, theologians, and professionals in health care and medicine from East Asia (China, Japan), Europe, and North America.
The first part of the book probes the foundations of personhood. Examining critically the main theories on personhood in contemporary philosophy, the authors offer alternatives that better respond to contemporary challenges and their implications for bioethics.
The focus of the second part is firmly on the Confucian relational concept of the person and on the social constitution of personhood in traditional Japanese culture. While the essays challenge the individualistic features of personhood in the Western tradition, they lay the foundations for a richer concept that holds great promise for the resolution of moral dilemmas in modern medicine and health care.
The third part of the book enters into a dialogue with the Christian tradition and draws on its spiritual heritage in the search for answers to the contemporary challenges to human dignity and value. Its focus is on the Catholic social thought and Lutheran theology.
The fourth part addresses the moral status of persons in view of specific issues such as the effects of brain injury, gene therapy, and human cloning on personhood. It extends the scope of research beyond human beings and inquires also into the moral status of animals.

--------------------------------   *Editions Rodopi B.V.*

USA/Canada: 6075 Roswell Rd., Ste. 219, Atlanta, GA 30328, Tel. (404) 843-4314, *Call toll-free* (U.S.only) 1-800-225-3998, Fax (404) 843-4315

All Other Countries: Tijnmuiden 7, 1046 AK Amsterdam, The Netherlands. Tel. ++ 31 (0)20 6114821, Fax ++ 31 (0)20 4472979
orders-queries@rodopi.nl ----- http://www.rodopi.nl

ASPECTS IN CONTEXTS - STUDIES IN THE HISTORY OF PSYCHOLOGY OF RELIGION

Ed. by Jacob A. Belzen

Amsterdam/Atlanta, GA 2000. 299 pp.
(International Studies in the Psychology of Religion 9)
ISBN: 90-420-1521-7   Bound Hfl. 150,-/US-$ 64.-
ISBN: 90-420-1511-X   Paper Hfl. 60,-/US-$ 26.-

Psychology of religion has been enjoying considerable attention of late; the number of publications and people involved in the field is rapidly increasing. It is, however, one of the oldest branches within psychology in general, and one of the few in which an interdisciplinary approach has been kept alive and fostered. The fate of the field has been quite varied in the countries where psychology of religion has been initiated and developed during the 20th century. In this volume, some aspects of this international history are examined. Coming from six different Western countries, each of the contributors has a record in the historiography of psychology and profound knowledge of psychology of religion. Their approaches combine elements from the history of mentalities, the social history of science and biographical studies.

The volume contains in-depth treatments of such topics as the growth of the field as reflected in university politics, developments within international organizations, and the personal involvement of contributors to the field. A wealth of information is provided on the background of the work of well known psychologists of religion like James Henry Leuba, Oskar Pfister, Gordon Allport, Werner Gruehn, Antoine Vergote and others.

The editor, Jacob A. Belzen, has earned doctorates in Social Sciences, History, and Philosophy. He is a Professor of Psychology of Religion at the University of Amsterdam and the University of Utrecht (The Netherlands). He has published widely on the psychology of religion, especially on its history and foundations.

--------------------------------   *Editions Rodopi B.V.*

USA/Canada: 6075 Roswell Rd., Ste. 219, Atlanta, GA 30328, Tel. (404) 843-4314, *Call toll-free* (U.S.only) 1-800-225-3998, Fax (404) 843-4315

All Other Countries: Tijnmuiden 7, 1046 AK Amsterdam, The Netherlands. Tel. ++ 31 (0)20 6114821, Fax ++ 31 (0)20 4472979